Monographs in Electrical and Electronic Engineering

Editors: P. HAMMOND, D. WALSH

Monographs in Electrical and Electronic Engineering

R. L. Bell: *Negative electron affinity devices* (1973)
G. Gibbons: *Avalanche-diode microwave oscillators* (1973)

Infrared techniques

H. C. WRIGHT

CLARENDON PRESS · OXFORD

Oxford University Press, Ely House, London W. 1

GLASGOW   NEW YORK   TORONTO   MELBOURNE   WELLINGTON
CAPE TOWN   IBADAN   NAIROBI   DAR ES SALAAM   LUSAKA   ADDIS ABABA
DELHI   BOMBAY   CALCUTTA   MADRAS   KARACHI   LAHORE   DACCA
KUALA LUMPUR   SINGAPORE   HONG KONG   TOKYO

© OXFORD UNIVERSITY PRESS 1973

PRINTED IN GREAT BRITAIN
AT THE PITMAN PRESS, BATH

TO JEAN

Preface

An attempt is made in this book to adumbrate most topics that are useful and active in infrared practice today. The subject matter deals with infrared radiation from its generation, through various manipulations, to the final detection and display. Underlying mechanisms are stressed and related to the phenomenon or device behaviour under consideration.

Clarity has been sought before rigour in the simple analyses offered and these are supported with references to both standard and less commonly quoted literature.

During preparation I have drawn on the learning of the many associates that several decades in the electronics industry have afforded me. I would like here to acknowledge my indebtedness to them for their generous and patient deployment of scholarship.

My thanks are due to Mr. D. V. Eddolls for undertaking and discharging the laborious task of proof reading and to Dr. D. Walsh for editorial assistance throughout.

Finally, I must claim residual errors as my own work.

H. C. Wright

Allen Clark Research Centre,
Caswell, Towcester
February, 1973

Contents

1. INTRODUCTION ... 1
   1.1 Definition of the infrared spectrum ... 1
   1.2 Herschel's pioneering work ... 1
   1.3 Exploitation of infrared radiation ... 2

2. GENERATION, MANIPULATION AND MONOCHROMATION ... 5
   2.1 Introduction to infrared components and sources ... 5
   2.2 Infrared sources ... 5
   2.3 Infrared windows and lenses ... 10
   2.4 Reflecting elements ... 14
   2.5 Filters and monochromation ... 15
   2.6 Polarization in the infrared ... 20
   2.7 Absorbtion of infrared radiation on thermal detectors ... 20

3. VACUUM AND SOLID-STATE ELECTRONIC DETECTORS ... 22
   3.1 Electronic photodetectors ... 22
   3.2 The vacuum photocell ... 24
   3.3 The semiconductor photodetector ... 26
      (a) General considerations ... 26
      (b) Photoconductive cells ... 28
      (c) Photovoltaic cells ... 31
      (d) Photo-electro-magnetic cells ... 33

4. THERMAL DETECTORS ... 36
   4.1 Detection by thermal effects ... 36
   4.2 The thermal equation ... 38
   4.3 The Golay detector ... 39
   4.4 The bolometer ... 41
   4.5 Thermojunction and thermopile ... 43
   4.6 Pyroelectric detector ... 44

5. LESS ORTHODOX DETECTORS ... 47
   5.1 The unconventional detector and indicator ... 47
   5.2 The evaporagraph or eidophor ... 47

Contents

5.3	The absorption-edge image converter	48
5.4	Liquid crystals	49
5.5	Microwave-biassed photoconductors	50
5.6	The Josephson-junction detector	51
5.7	The electronic bolometer	52
5.8	Magnetically tuned infrared detector	53
5.9	The photon-drag detector	53
5.10	M.O.M. detector	54

6. NOISE, SIGNAL PROCESSING AND COEFFICIENTS OF PERFORMANCE   56

6.1	General aspects of noise and signal processing	56
6.2	Background noise in thermal detectors	61
6.3	Background noise for electronic detectors	65
6.4	Johnson or Nyquist noise	66
6.5	Current noise	67
6.6	The signal-to-noise ratio	67
6.7	Heterodyne detection	68
6.8	Coefficients of performance for detectors	70

FURTHER READING   73

INDEX   77

List of symbols used

- A cross-sectional area
- α coefficient of restoring force on Golay diaphragm
- B magnetic induction
- β temperature coefficient of resistance
- C electrical capacity
- c velocity of light
- D diffusion coefficient
- η quantum efficiency
- ϵ dielectric constant
- E photon energy
- $\mathcal{E}$ electric field
- ϕ work function
- F photon incidence rate
- f frequency
- Γ pyroelectric coefficient
- I_s signal current
- J refractive index
- k Boltzmann's constant
- K thermal conductivity
- λ wavelength
- L^* diffusion length
- μ charge-carrier mobility
- m avalanche multiplication factor
- n density of free electrons
- N total number of free electrons
- ν Stefan's constant
- ψ radiant power
- P photon flux
- p photon momentum
- q charge on electron
- R electrical resistance
- $\mathcal{R}$ thermal resistance
- σ electrical conductivity
- Σ absorption coefficient
- s volume specific heat
- $\mathscr{S}$ thermoelectric coefficient
- τ response time, carrier lifetime
- τ_R dielectric relaxation time
- t time
- θ temperature
- V applied voltage
- V_S signal voltage

1 Introduction

1.1. Definition of the infrared spectrum

The average human eye is most sensitive to electromagnetic radiation of wavelength about 0·55 μm and, as the wavelength increases, the eye's response decreases until at about 0·7 μm it can no longer detect any radiation. This latter wavelength therefore forms a good definition for the beginning of the infrared (i.r.) spectrum. The strict lexicographer would perhaps not allow any upper limit to the region, but in practice it is usual to regard wavelengths longer than 100 μm as beyond even the far infrared. This is not entirely arbitrary, since radiation of longer wavelength is normally generated by bulk electronic oscillations rather than the energetic transitions of individual electrons that produce visible and infrared radiation (Fig. 1.1).

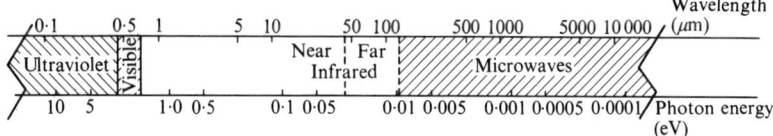

Fig. 1.1. Position of infrared in the electromagnetic radiation spectrum.

The spectral range 0·7–100 μm, representing photon energies of 1·55–0·012 electronvolts, is so large as to require subdivision, and is broken down into the near infrared and the far infrared, with a rough boundary at 50 μm. We shall discuss generation, manipulation and detection for radiation lying in each of these two regions.

1.2. Herschel's pioneering work

It may be of interest to consider briefly the history of the subject: any man who had enjoyed the warmth of the sun must have had some concept of the nature of i.r. radiation, but the organized investigation of the subject started only with William Herschel. This young German military musician, while visiting England with his regiment, apparently liked it well enough to abandon Army life and remain in the country earning his living as a concert musician, while still pursuing astronomy as a hobby. Dedication and good fortune brought him the reward of a

2 Introduction

civil list pension from George III when he discovered the planet Uranus in 1781. Free now to follow his scientific interests, he engaged in solar spectroscopy. While so occupied, Herschel observed that thermometers placed just beyond the visible region of a spectrum he had produced in his laboratory showed a marked increase in temperature. This was enough for him rapidly to establish the existence of an invisible radiation obeying laws of refraction and reflection similar to those obtaining for visible light. A thermometer, possibly with a blackened bulb to enhance i.r. absorption, may seem a naive instrument when considered in the context of the many and sophisticated detectors we shall discuss later, but it is salutary to remember that the present-day Golay detector, having a sensitivity near the theoretical limit, is in effect a constant-volume gas thermometer.

1.3. Exploitation of infrared radiation

We must now consider whether the study of i.r., in addition to being of interest, is of practical use. Fortunately for the health of our chosen subject and the well-being of its practitioners, the answer is yes. We discuss here four areas in which i.r. radiation is exploited:

(1) analysis and characterization of materials
(2) thermal imaging
(3) communication
(4) astronomy.

(1) Analysis and characterization of materials

Certain organic molecular groups resonate at well-defined frequencies when stimulated with radiation in the i.r. spectrum. These resonances are made manifest by the strong absorption of particular wavelengths in solutions containing the atomic groupings under investigation. These absorption lines are so precise and well documented that they form a most powerful tool in organic analysis. Another analytic use lies in the excitation of electrons between various energetic levels in semiconducting materials. A photon of the appropriate energy may be absorbed by an electron lying in one energy level which will then transfer to a second level of higher energy. This action is made evident by a change in the electrical conductivity of the material and by this means a map of the various energetic states can by made. These are amongst the most important characteristics of a semiconductor. It is evident that this effect could also be used to detect the presence of i.r. radiation and, indeed, it does form the basis of a large group of practical detectors.

The existence of impurities in a crystalline lattice can be shown up by, again, an i.r. absorption pattern. Any impurity or defect at a lattice site will have some natural resonance frequency, which is superimposed on the fundamental lattice vibration spectrum, and it will absorb any radiation having this frequency. The advantage of using the lattice-defect resonance rather than the previously mentioned excitation between energy levels is that the defect need not be ionizable to be detected, and hence electrically inactive impurities may be recognized.

Introduction 3

Consider now the free charge-carriers in a semiconductor: these can be regarded as forming a plasma with its own resonant frequency. This frequency is determined by both the density and the effective mass of the carriers and will, if measured, give either parameter if the other be known. Observation of this plasma-frequency over the surface of a semiconducting crystal slice has been used to give a very precise plot of the carrier-density distribution.

(2) *Thermal imaging*

The practice of thermal imaging makes use of the fact that any material body emits electromagnetic radiation with a wavelength distribution characteristic of its absolute temperature and surface emissivity. Terrestrial bodies having temperatures in the neighbourhood of 300 K radiate most strongly at about 10 μm. Small variations in temperature over the surface of the body will, provided the emissivity is constant, form what is called a thermal image, by analogy with the optical image formed by reflected visible radiation. The thermal image is made visible in practice by scanning the body and measuring the intensity of radiation from each point. This information is then used to produce an optical image which reproduces the temperature intensity pattern. Such a thermal imaging system exploits most aspects of i.r. technology, involving, in contemporary manifestations, sophisticated imaging, detection, signal processing, and display techniques. The details in the human thermal pattern are used in medical diagnosis, classically for the early detection of breast cancer and circulatory disorders but also, perhaps most significantly, for determining the precise area of irretrievably lost flesh in burning accidents. Early knowledge of this is of value in making skin grafts of adequate but not excessive size.

Apparatus using electrical or mechanical power will normally show a pattern of elevated temperatures which can be used to assess the condition of the system and forecast possible approaching malfunctions. Power transmission lines are conveniently monitored for 'hot-spots' by i.r. cameras, and continuous strip production processes, whether paper, fabric, or metal, also require remote temperature measurement without physically connected thermometers. Patterns of cooling water, or effluent, in rivers are shown clearly by airborne thermal imaging, and some geological features may be located by differences in surface soil temperature.

Security is evidently served by any system which can see men and machines in the dark, solely by their own self-emitted thermal radiation, and civilian and military uses can both be envisaged.

(3) *Communication*

The advantages of using i.r. for communication can still be debated. It shows many of the advantages of normal optical techniques in that it can be generated and modulated by fairly simple apparatus compared with the complexity of a conventional wireless transmitter. Since it can be confined to a precisely directed small-area beam, it is more efficient in its use of power and, being invisible, it can be more easily secured against unauthorized detection should this aspect be important. We can add the absence of interference between different signals to the list of benefits, but it is necessary to set against these the limitation that no opaque bodies must lie on the line joining the source and detector. When it is remembered that a kilometre or two of the normal terrestrial atmosphere is an opaque body for some i.r. wavelengths, this becomes a serious restriction in its use, which is rendered even more significant by the rather low power of the sources which are available.

An application which does not easily fit into any general category, but is by no means the least interesting, is i.r. laser range-finding. The range-finder works by emitting a pulse of radiation which is reflected off a distant object and received back again at the instrument. The departure and return of the pulse are observed by a detector and the time interval between them converted into a distance much in the manner of a conventional radar system. Since the gauging of an object as far as one kilometre away would give a pulse separation of about 10^{-7} second, it is evident that the detector used must not only be sensitive, but have a very rapid response time.

(4) *Astronomy*

This chapter will end with a brief description of i.r. astronomy, the science which brings together two of Herschel's major interests. The three spectral areas which have been most used for examination of extra-terrestrial phenomena are the visible, the i.r., and the radio-frequency regions, since electromagnetic radiation other than that falling in one of these groups is strongly absorbed by the earth's atmosphere.

I.R. has been the last of the three to develop by reason of the low power of i.r. sources in space, excepting our own sun and, until recently, the insensitive radiation detectors available. Science and technology have substantially remedied the second cause in the past decade, and now that a well-filled map of astronomical i.r. sources is available, astronomers are turning from cataloguing to characterization of the sources.

If the source be a hot body then the relationship between radiation intensity and wavelength is fixed and characteristic of the body temperature. If this intensity –wavelength relationship is not satisfied then the radiation may have suffered reflection from a cold, quiescent body or, alternatively, may have originated in a cloud of interstellar electrons moving through a magnetic field with velocities near that of light. The polarization of the received radiation is a further guide to the nature of the source, and eventually one may hope to denote all observed sources as electron clouds or stars in their various periods of birth, death, or radiant life.

These examples of the practical uses to which i.r. can be put could be multiplied many times and still be less than exhaustive. They do, however, point out the regions which are being vigorously served by i.r. technology today, although one fears (hopes?) that any catalogue of this nature is constantly and rapidly being made obsolete.

2 Generation, manipulation, and monochromation

2.1. Introduction to infrared components and sources

It is felt that the most lively area in the study of i.r. is probably in the design and fabrication of the devices used for detecting the presence of radiation. This attitude is reflected in the amount of space which will be devoted in the book to detectors of various types, but in the present chapter we will be concerned only with the generation of radiation and its manipulation. A convenient and systematic way of considering the problems that may be met in any system using i.r. radiation is to suppose that it is necessary to generate, monochromate, refract, reflect, polarize, transmit, and then absorb the radiation. Each of these operations will be discussed in some detail (Fig. 2.1).

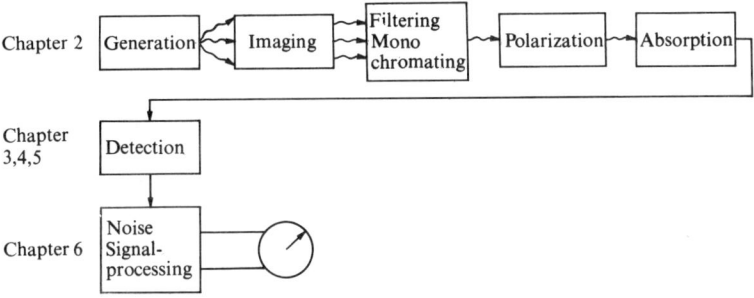

Fig. 2.1. Topic plan.

2.2. Infrared sources

The problems in generating infrared compared with visible radiation are those of inadequate power. The simple incandescent source is very inefficient when emission in the longer wavelengths is required. The energy of wavelength $\lambda \to (\lambda + d\lambda)$ radiated from a body at an absolute temperature T can be written as $\psi_\lambda d\lambda$ per unit area, where

$$\psi_\lambda = K_1 \lambda^{-5} (\exp(K_2/\lambda T) - 1)^{-1} \qquad (2.1)$$

and K_1 and K_2 are universal constants (Fig. 2.2). This relationship between energy and wavelength only holds when the heated body is 'black', the 'black body' being by definition one which absorbs all radiation falling on it, whatever the

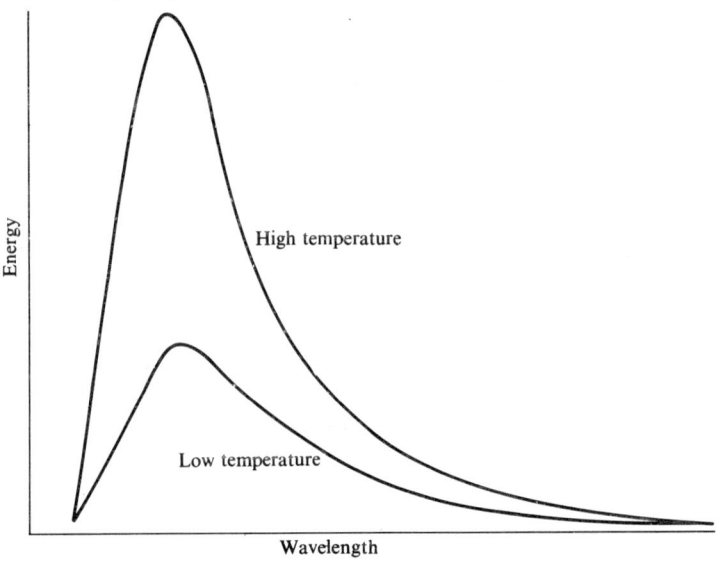

Fig. 2.2. Emission patterns of a black body.

wavelength. This precise emission pattern that black bodies display is useful for the calibration and testing of detectors, since the total energy in any wavelength range can be computed knowing the temperature and area of the emitting body. There are, however, two adverse results from the form of eqn (2.1). Firstly the energy is very small for large λ, i.e. in the i.r., as λ raised to the fifth power appears in the denominator. Secondly, the maximum power is emitted at a wavelength which decreases as the temperature is increased so that attempts to increase the emitted power by increasing the temperature of the incandescent body are defeated, or at least debilitated, by transfer of the peak power emission to a shorter wavelength.

The factors with which it is necessary to contend are made plain if it is assumed that a limited electrical power W watts is available to provide an incandescent source of $\lambda \to (\lambda + d\lambda)$ wavelength radiation. The only apparent variable in the hands of the designer is the surface area A of the incandescent body. To obtain T_B the absolute temperature of the body we use

electrical power into body = radiated power lost
$$W = A\nu (T_B^4 - T_A^4), \tag{2.2}$$

where ν = Stefan's constant and T_A = ambient temperature.

This gives $T_B = [J/A\nu + T_A^4]^{1/4}$.

The power emitted from the body in the $\lambda - \lambda + d\lambda$ range is then, by reference to eqn (2.1),

$$\psi_\lambda d\lambda A = AK_1\lambda^{-5} \left(\exp \frac{K_2}{\lambda(W)/A\nu + T_A^4)^{1/4}} - 1 \right)^{-1} \tag{2.3}$$

Generation, manipulation and monochromation

This expression has a maximum value of approximately

$$4K_1 W/\lambda \nu K_2^4$$

when $A = 16\lambda^4 W/K_2^4 \nu$.

It is thus seen from the first expression that the available radiation from a given electrical power decreases inversely as the wavelength increases and, from the second expression, that even this requires the optimum area of the incandescent body to increase as the fourth power of the wavelength. If then we had a filament lamp efficiently providing 0·5 μm visible radiation and required 10 μm i.r., it would be necessary to increase the filament area by over 10^5 times and it would then offer only 1/20th of the radiated power. These magnitudes are obviously not practicable and we need to depart from the black-body radiation spectrum defined in eqn (2.1). It is found that the radiation from silicon carbide or thorium oxide is concentrated strongly in the longer wavelengths, or, to rephrase this, the emissivity is higher for i.r. than for visible radiation in these materials and the black-body spectrum does not obtain.

In use a short rod, either made from silicon carbide or by coating platinum with thorium oxide, is heated by passing an electric current through it and such a globar or Nernst filament can make a satisfactory source for radiation of wavelengths up to about 100 μm.

The region beyond 100 μm is served by the emission from gases which have been raised to a temperature where the constituent atoms are mostly ionized, i.e. electrons have been removed from constituent atoms so that they exist as plasmas—interpenetrating clouds of positive ions, electrons, and the remaining unionized atoms. The collisions between these particles give rise to a non-black-body spectrum rich in longer wavelength radiation up to more than 300 μm. The most widely used plasma source is the mercury arc, in which an electrical discharge is effected through mercury vapour at a pressure of several atmospheres, raising it to a temperature of about 6000K.

The black-body, globar, and plasma sources all give a continuous spectrum of radiation, from which a particular wavelength can be selected by use of the monochromation techniques discussed in the next section. If a source giving only a single wavelength emission is adequate, however, the powerful laser can be used. 'Laser' (**L**ight **A**mplification by **S**timulated **E**mission of **R**adiation) is obviously a misnomer as we are concerned with i.r. radiation and not light, but the acronym is now entrenched in the literature and we shall not attempt reformation. The fundamental difference between lasers and other i.r. sources lies in the manner in which each emitted photon is produced. In an incandescent source packets of energy are stored in the radiating body and every now and then the stored energy is released spontaneously with the emission of a photon. These haphazard events yield photons in irregular bursts of varying energies travelling in random directions. In contrast the laser is made from material selected to have two well-defined energy levels in which electrons may exist, the difference in energies being equal

8 Generation, manipulation and monochromation

to that energy we require the emitted photon to have. Electrons are then 'pumped' up into the higher level where they are stored. This pumping can be done either by bombarding the material with photons from another, non-laser, source or by injecting high-energy electrons into the material. Now a single photon is sent through the laser and as it passes each excited electron it triggers the collapse to the lower energy level with the emission of a photon. The significance of this is that the stimulated photons all travel in the same direction and have the same phase as the triggering photon. Overall we have collected and stored photon energy from the pumping source and released it in a continuous coherent stream of radiation all travelling in one direction (Fig. 2.3). To appreciate the potency

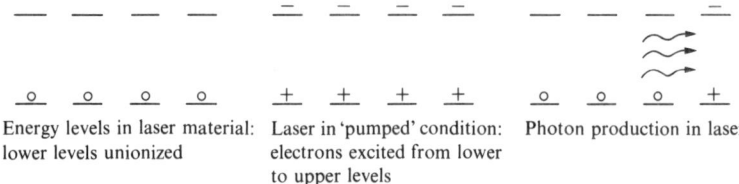

Energy levels in laser material: lower levels unionized   Laser in 'pumped' condition: electrons excited from lower to upper levels   Photon production in laser

Fig. 2.3. Schematic of laser mechanism.

of this operation consider a normal radiant source of 1 watt. The intensity 0·1 m away will be $(1/4\pi)10^2$, about 10 W m^{-2}. A laser of similar power might emit a beam of radiation with a cross-section of 10^{-6} m^2, giving an intensity of 10^6 W m^{-2}. Laser material can be found in the solid, liquid, or gaseous states, with one of the most interesting systems being the solid diode-laser. Here the pumping is done very elegantly by forcing electrons over a potential step built into a bar of semiconductor. These are then stored in a high-energy condition until triggered by a photon in the usual way. Table 2.1 shows some of the available laser wavelengths for use in the i.r.

For the next increase in radiation wavelength it is necessary to leave optical practice and move to the extreme limits of microwave generation. When a conventional thermionic valve oscillates the variations in current are caused by changes

TABLE 2.1
Laser sources

Principal operating wavelengths (μm)	Material	Type
0·9	GaAs	Solid (Diode laser)
3·8	InAs	Solid (Diode laser)
8–28 (variable with the Pb/Sn ratio in the alloy)	(Pb–Sn) Te	Solid (Diode laser)
10·6	CO_2	Gaseous
47·7, 78·5, 120	H_2O	Gaseous
95·8, 216	He	Gaseous
310, 336	HCN	Gaseous
538	ICN	Gaseous

Generation, manipulation and monochromation

in the density of the electron stream flowing from cathode to anode. The rate at which the total current can be changed or modulated will be limited by the finite inertia of the electrons, and in order to reach microwave frequencies the technique of velocity modulation is used. In this electrons are grouped together in bunches without altering the total current through the valve. These electron bunches pass between grids connected to a resonant circuit. As each bunch passes, a current pulse is induced and oscillations are generated, having a period equal to the time between the arrival of successive bunches.

This is the principle of the Klystron valve, which is shown schematically in Fig. 2.4. Electrons leave the cathode C and travel towards the anode A. An alternating potential applied between the 'bunching' grids BB' successively retards and accelerates the electrons towards the 'catcher' grids CC' causing them to form the required bunches bb. (The complex mechanism of bunching is discussed more exactly in Parker (1958).)

As the bunches pass CC' a pulse of current is induced in the resonant external circuit and coupled back to a similar circuit connected to BB'. This coupling, or feed-back, causes the valve to oscillate and frequencies corresponding to 0·5 cm wavelength radiation can be generated. To obtain the shorter wavelength required their electric field is applied to a non-ohmic resistor. In this type of resistor the current I flowing is not proportional to the first power of the voltage V as in a simple resistor, but depends on second or even higher powers:

$$I = AV + BV^2 + \ldots, \tag{2.4}$$

$A, B,$ being constants.
If the applied voltage varies with a frequency f:

$$V = V_0 \cos 2\pi ft, \tag{2.5}$$

then $I = \tfrac{1}{2} BV_0^2 + AV_0 \cos 2\pi ft + \tfrac{1}{2} BV_0^2 \cos 4\pi ft + \ldots$. $\tag{2.6}$

One of the most effective non-ohmic resistances is formed by a high-pressure argon plasma. This frequency-multiplying, if carried far enough, will eventually give wavelengths as short as 100 μm, but the method is inefficient and only small radiated powers can be obtained. This has led to the development of efficient but expensive devices known as backward-wave oscillators or carcinotrons. In this source high-velocity electrons, constrained by a magnetic field to remain in a narrow beam, are passed through a periodic comb-like metal structure. The passage of the electron charge produces a fluctuating field in the 'comb', the fundamental frequency of which is equal to the electron velocity divided by the periodic distance between 'comb teeth'. One of the harmonics of this fundamental actually travels backwards, i.e. in a direction opposite to the electron flow, and returns energy to the electron-emitting end of the tube, causing the device to oscillate. This energy-return system can be compared with the manner

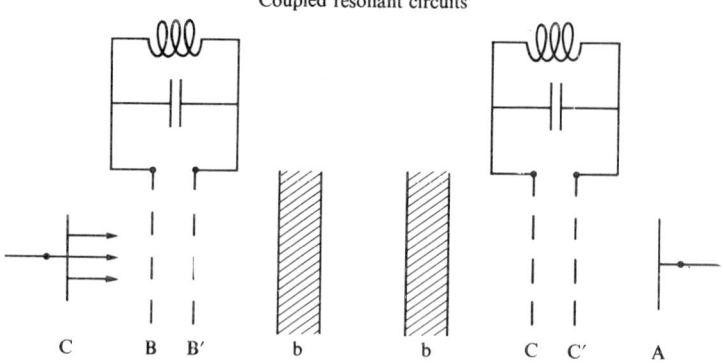

Fig. 2.4. Schematic of Klystron valve.

in which a simple triode valve can be made to oscillate when the anode is connected externally to the grid, the connection in the carcinotron being made internally by the vehicle of an electromagnetic wave. One watt of 1000 μm radiation has been obtained with this technique.

2.3. Infrared windows and lenses

The simple glass refracting components of the optical system will only serve for a very limited region in the i.r., as radiation with wavelengths over 3 μm will be heavily absorbed. There are three phenomena which may prevent a given material from being used successfully to make a lens or window in an i.r. system. Firstly the reflectivity may be so great that only a small fraction of radiation power actually enters the component. The reflection at an abrupt interface between two media depends on the refractive indices J_1, J_2 of the media concerned.

$$\text{Reflectivity} = (1 - J_1/J_2)^2/(1 + (J_1/J_2)^2). \tag{2.7}$$

If a semiconductor is to be used as a window or lens then, because semiconductors have in general J values of about 4, the interface between semiconductor and air, with its J of nearly 1, will have a reflectivity of about 1/3. Without remedy this would make any semiconducting system-element very lossy since, even without subsequent loss by absorption, only 2/3 of the incident radiation would pass through. It is possible in principle to eliminate this loss entirely by grading the refractive index smoothly from J_1 to J_2 and so avoiding the presence of a sharp interface. Complete gradation is not generally feasible in practice but the application of one layer of refractive material at the interface will reduce the reflection to a few per cent for one specified radiation wavelength.

The layer thickness t and refractive index J_L must satisfy

$$J_L = \sqrt{(J_1 J_2)}$$
$$t = \lambda/4$$

Generation, manipulation and monochromation

where λ is the radiation wavelength in the layer material. The reflection is then inhibited by interference between radiation reflected back from the film–semiconductor interface and that just entering the air–film interface. These two radiation trains will at all times add to zero at the air–film interface if the quoted requirements are satisfied (Fig. 2.5). The application of these anti-reflection coatings involves the controlled deposition of a dielectric material having the appropriate refractive index to a precision of perhaps 0·1 μm.

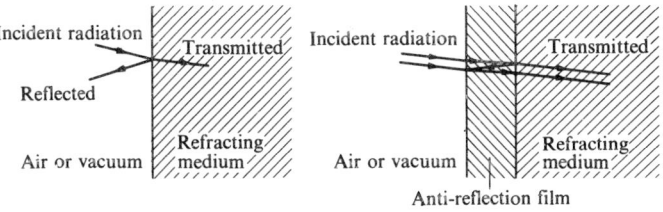

Fig. 2.5. Effect of anti-reflection coating.

Having overcome the problem of reflection either by coating, or selecting material with a *J* value of nearly one, absorption within the material must be considered. There are two basic loss mechanisms in the passage of radiation through a medium: one involves the cyclic movement of charged particles in the material caused by the alternating electric field of the radiation, and the other depends on the liberation of charged particles by the absorption of photons. The characteristics of the absorption are, of course, a consequence of the loss mechanism which is involved and we will examine these mechanisms under five sub-groups.

(*a*) *The excitation of bound orbital electrons from the atoms in a semiconductor into a free, mobile, state.* The energy necessary to do this is called the band-gap energy, E_g, and a photon having less energy than E_g will pass through without absorption, those with energy greater than E_g being almost wholly absorbed. We thus have a medium which changes very abruptly from being transparent at long wavelengths to being completely opaque at short wavelengths. Some semiconductors with their transition wavelengths are shown in Table 2.2.

TABLE 2.2
Transparency regions for semiconductors

Semiconductor	Transition wavelength (μm)
GaAs	0·9
Si	1·2
Ge	2·0
InAs	3·8
InSb	7·7
Pb–Sn Te	10–30
	(Dependent on Pb/Sn ratio)

12 Generation, manipulation and monochromation

(*b*) *The excitation of bound electrons from, or into, an impurity atom in a semiconductor lattice.* Unlike the processes in (*a*) the absorption is only effective when the photon energy is approximately equal to that required by the excitation process. As it is essentially a smaller energy than (*a*) it will be manifest as a broad absorption peak in the low-energy transparent region that the pure semiconductor shows.

(*c*) *The oscillation of free carriers.* If there are many free carriers in the semiconductor then at low photon energies, and hence long-wavelength radiation, they may be excited into alternating current, with its usual ohmic energy loss, and hence become a radiation-absorbing mechanism. This shows as a slow loss of transparency in the wavelength region beyond the impurity absorption peaks, and can be represented analytically by an absorption coefficient Σ related to the dielectric constant ϵ and σ the electrical conductivity of the free carriers, by

$$\Sigma = 3 \times 10^2 \times \sigma/\epsilon \text{ m}^{-1}. \qquad (2.8)$$

(*d*) *Induced vibrations of the lattice.* A crystal lattice may be regarded as possessing some fraction of a complete electron charge located at each of its lattice sites. The fraction ranges from zero, for a completely non-ionic material, to unity, for the completely ionic compound. The charged sites will be set into oscillation by the high frequency i.r. radiation and in this manner energy will be absorbed from the radiation. This loss mechanism obtains even in glassy material as the constituents here are still sufficiently ordered in space, over small volumes, to behave as lattices. Absorption is not avoided entirely in the completely non-ionic material, as the molecules present may be cyclically deformed (i.e. polarized) by the radiation field, and absorb energy in this way.

The absorptions discussed above are generally referred to as fundamental or lattice absorptions. Unlike the semiconductor transmission spectrum these mechanisms give good transmission below some critical wavelength and then, for longer-wavelength radiation, become opaque. Table 2.3 shows a number of

TABLE 2.3
Transparency regions for dielectrics or non-conductors

Material	Approximate long-wavelength limit of transparency (μm)
Quartz	4·5
Magnesium oxide	8·5
'Irtran 1' (Proprietary i.r. transmitting glass)	8·5
'Irtran 2'	14·5
Sodium fluoride	15
Arsenic–selenium glass	18
Sodium chloride	26
Thallium bromide	40
Diamond	80

Generation, manipulation and monochromation 13

crystalline non-conducting media. As these generally have a low dielectric constant, and hence a low reflectivity, the provision of anti-reflection coatings is less important, but, to set against this, if the material is needed for an i.r. lens, the smaller refractivity is a disadvantage (Fig. 2.6).

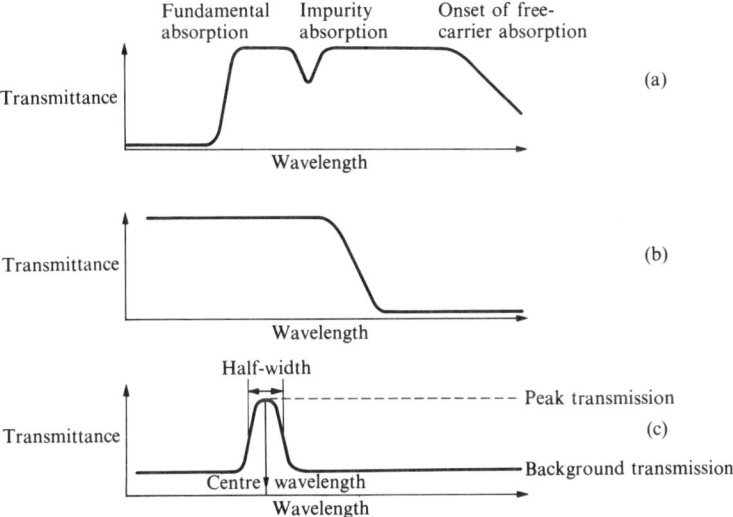

Fig. 2.6. Absorption patterns. (a) Semiconductor. (b) Dielectric. (c) Band-pass filter.

(*e*) *Characteristic or molecular vibrations.* It was mentioned in the first chapter that certain atomic groupings will resonate with a characteristic frequency, often falling in the i.r. spectrum, which allows them to be identified as constituents of an unknown material by spectroscopy. These resonances give limited regions of opacity known as absorption bands, and it is this effect which may prohibit the use of plastic as window material in a given i.r. system, despite the fact that many organic plastics can be obtained in the form of very thin but comparatively strong sheeting, an otherwise very convenient and easily worked material.

Some organic materials and their characteristics are shown in Table 2.4.

One medium of transmission which must be carefully considered in any terrestrial system is the atmosphere itself. Impurities present, particularly water

TABLE 2.4
Absorption bands in plastic materials

Material	Absorption bands (μm)
Polyethylene	3, 7, 14
Polymethylmethacrylate	3, 6, 7, 8, 9, 10, 12, 14
'Cellophane'	3–4, 6
(Proprietary cellulose film)	

14   Generation, manipulation and monochromation

vapour and CO_2, give rise to a complex relationship between radiation wavelength and atmospheric absorption. The situation is simplified, in concept, by regarding the limited regions of the spectrum which show high transparency as 'windows', and the diagram indicates the significant windows in the i.r. region (Fig. 2.7).

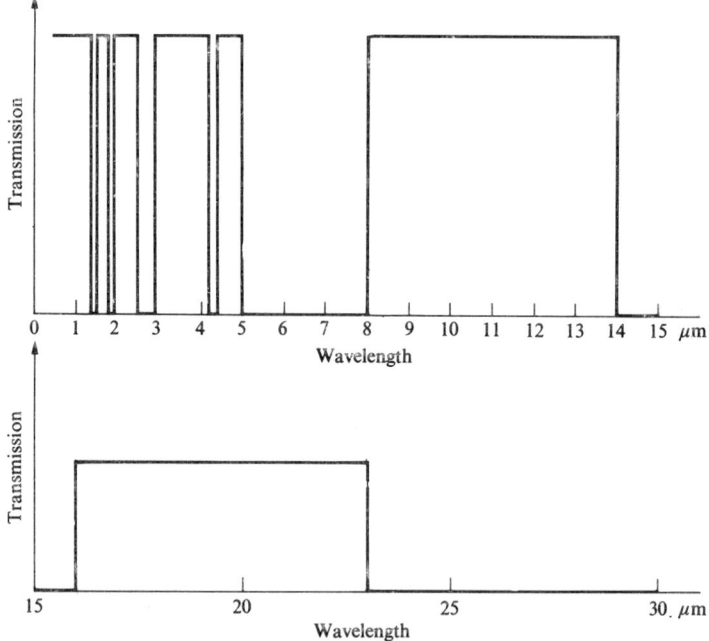

Fig. 2.7. Wavelength regions of high transmittance in the atmosphere.

2.4. Reflecting elements

The problems encountered in making suitable refracting elements or lenses for use in the i.r. will be avoided if it is possible to use reflecting elements for imaging and focusing. As the permissible surface irregularity in an optical mirror is proportional to the wavelength of the radiation in use there is an obvious relaxation of the tolerances where they are to be used for the infrared. They are usually made by the deposition of a thin metallic film on a suitably shaped substrate and, provided its geometry is stable, the substrate material is not too important. The metal film may, however, show a considerably different reflectivity to i.r. radiation from that which it displays to light. The method used to deposit the metal film is very important and, with optimum preparation conditions, a silver film will show a reflectivity of 0·99 for wavelengths greater than 0·6 μm. Gold mirrors have greater chemical stability but do not achieve the 0·99 reflectivity for wavelengths less than 2 μm, while aluminium shows poor reflectivity until 30 μm and indeed has a reflection minimum at 0·8 μm. This is shown in Fig. 2.8. The value of 0·99 has

Generation, manipulation and monochromation 15

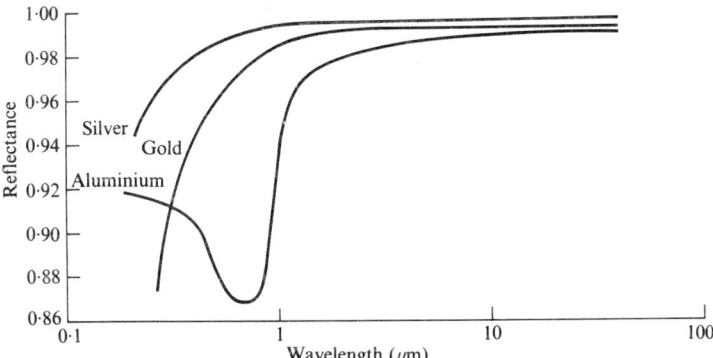

Fig. 2.8. Reflection in the infrared (Bennett and Ashley 1965)

been taken, arbitrarily, as an acceptable reflectivity. Such a value means that a system in which the imaging and focusing involved five reflections, not on this count a very sophisiticated apparatus, would lose 5 per cent of radiation energy.

2.5. Filters and monochromation

The need to restrict the range of wavelength which may be admitted to any system can be met with varying degrees of sophistication. By using a high-pass filter (H.P.F.) all radiation with a wavelength shorter than some pre-assigned value will be admitted and the dielectric transmitting materials, such as those listed in Table 2.3, can be used directly for this type of filter. To prevent radiation with a wavelength shorter than the pre-assigned value entering the apparatus a low-pass filter (L.P.F.) made from a semiconducting material such as one of those listed in Table 2.2 may be used.

An arrangement which can be made to provide either a L.P.F. or H.P.F. makes use of the phenomenon known as frustrated total-internal-reflection. If radiation A, Fig. 2.9(a), strikes the face of a prism at an angle greater than $\sin^{-1}(1/J)$, where

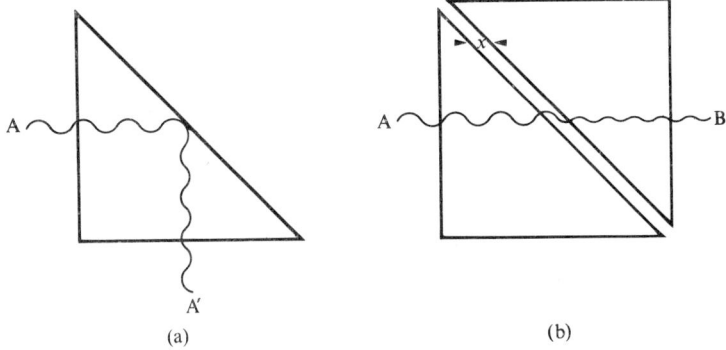

Fig. 2.9. Frustrated total-internal-reflection filter.

16 Generation, manipulation and monochromation

J is the refractive index of the prism material, then the ray is reflected back into the prism. There is however a small disturbance propagated beyond the prism face. This disturbance decays rapidly in space unless a second prism is brought into close proximity as in Fig. 2.9(b). The radiative disturbance will then travel in the second prism without further diminution. The separation x between the two prisms must be less than about half the radiation wavelength for this 'leak' to take place, and hence radiation of wavelength greater than $2x$ will pass through to B while that of shorter wavelength will suffer total-internal-reflection to A′. Selection of the appropriate emergent beam will make the device either a H.P.F. or L.P.F.

A filter that cuts off radiation above one assigned wavelength and at the same time cuts off radiation below a second, longer, wavelength forms a band-pass filter (B.P.F.). One obvious way of making a filter of this nature is to use both a H.P.F. and L.P.F. consecutively in the radiation beam. A more elegant B.P.F., the Christiannsen filter, exploits the variation in refractive index with wavelength shown by most transparent materials. If particles of a solid are suspended in a fluid matrix they will act as scattering centres for all radiation wavelengths except those for which the refractive indices of both particle and matrix are the same. Here the particles will 'disappear' and the filter transmit fully. The behaviour of the refractive index is also utilized in another B.P.F. which is particularly useful in the middle ranges of the infrared spectrum. Certain materials show an anomalously high refractive index at a particular wavelength. This enhances the reflectivity at this wavelength and if polychromatic radiation is made to suffer a succession of reflections from such a material this wavelength will become dominant. (See Table 2.5.)

TABLE 2.5
Anomalous reflection wavelengths

Material	High-reflectivity wavelength (μm)
ZnS	25
GaAs	35
InSb	55
KBr	85

The Fabry–Perot filter is the basis of a very flexible B.P.F. which can, by careful manufacture, produce a precisely shaped and located band-pass. In its simplest form the Fabry–Perot interference filter consists of two partially reflecting surfaces separated by a thin intermediate layer of transparent material as shown in Fig. 2.10(a). If polychromatic radiation is incident normally on the filter it will undergo multiple reflections as shown and the transmitted radiation will be reinforced by constructive interference when the effective separation w and the radiation wavelength λ satisfy the equation

$$2w = r\lambda, \qquad (2.9)$$

Generation, manipulation and monochromation 17

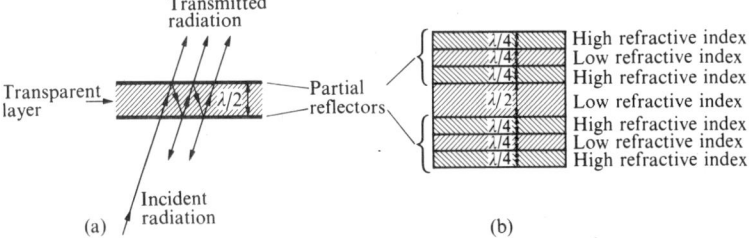

Fig. 2.10. The Fabry–Perot interference filter.

where r is an integer. This means that the filter will pass bands of radiation having wavelengths centred on $2w$, w, $2w/3$, ... The width of those bands becomes smaller with increasing reflectivity of the surfaces, but if a simple metal-film reflector is used the power absorbed in the metal will make the overall efficiency of the filter low and little radiation will be transmitted. Because of this, advantage is taken of the reflectivity consequent on an abrupt change of refractive index and the surfaces are formed from two layers of, respectively, high and low refractive index. The system is further refined by making each reflecting element from a stack of high/low refractive index plates a quarter of a wavelength thick, so that constructive interference occurs in the reflected beam. This makes a highly reflecting low-loss element, thus reducing the band-width without diminishing the overall radiation transmission by absorption in thick metal films.

Fig. 2.10(b) shows the final complexity achieved for a high-efficiency narrow band-pass filter. The work by O. S. Heavens (1955) gives a detailed description of these filters.

For radiation with a wavelength much greater than 50 μm we again borrow from microwave practice. A very simple resonant circuit can be made for microwaves with their frequencies of $\sim 10^{10}$ Hz by cutting a rectangular slot in a metal plate. The two regions of the plate having edges perpendicular to the electric vector in the radiation behave like an air dielectric capacitor (Fig. 2.11), while those perpendicular to the magnetic vector show inductance by virtue of the induced eddy currents in the metal. Possessing both inductance and capacitance the component, called an iris, will thus have an inherent resonant frequency.

To make an iris resonant at infrared frequencies of $\sim 10^{13}$ Hz the construction takes the form of 1 μm thick copper crosses positioned on a 2·5 μm-thick P.T.F.E. sheet in a grid pattern but electrically isolated one from another. Several sheets are placed one behind the other in the path of the radiation with some 10 μm separation between sheets. The resonant components thus formed absorb selected regions of infrared wavelengths allowing H.P.F., L.P.F., and B.P.F. to be formed. Ulrich (1968) describes the detailed construction and performance of these filters.

B.P.F.s are characterized by their background transmission (ideally zero); peak transmission (ideally 100 per cent); centre wavelength, i.e. the wavelength at the

18 Generation, manipulation and monochromation

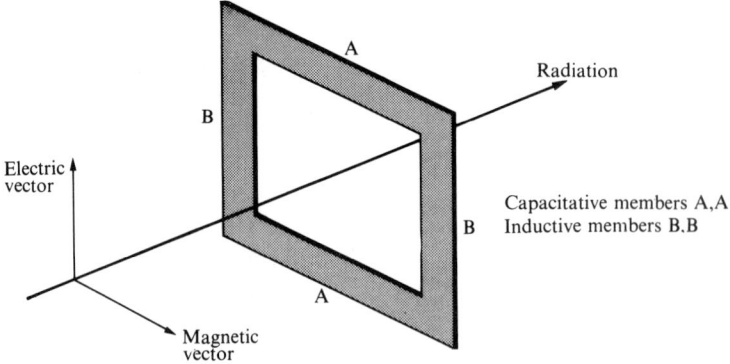

Fig. 2.11. Microwave iris.

centre of the band of transmitted radiation, and half-width, which is the difference between the two wavelengths at which the filter shows half of its maximum transmission. A typical commercial filter might show a peak transmission of 75 per cent at 10 μm with a half-width of 1 μm (Fig. 2.6).

Finally, where filtering is too coarse a technique, we must consider true monochromation. The problems of available radiation power are even worse now as the fraction of available flux that is used obviously becomes smaller as the monochromation becomes better, i.e. the band pass becomes smaller. For this reason, although the design principles of visible-radiation monochromators still obtain in the infrared provided the optical components refract and reflect effectively, the prism monochromator with its long optical paths through refracting material is usually rejected in favour of a reflecting-grating instrument in i.r. work. Similar in principle to the grating instruments used for the visible region, it has a practical advantage in that the gratings can be more coarsely ruled for the longer wavelengths. The simple substitution of a suitably ruled grating in an optical monochromator will suffice for wavelengths up to 15–20 μm, after which it is necessary to use an evacuated instrument to avoid atmospheric absorption. As well as the normal precautions taken against unwanted or stray radiation in monochromators it is necessary in i.r. work to remember that the entire instrument will be emitting radiation appropriate to a body at a temperature of about 300K. Some of this radiation, peaking at about 10 μm, will inevitably fall on the detector but, if the radiation beam is interrupted periodically, it will be possible to distinguish between the varying response to this and the steady background from the monochromator body. The evacuated grating monochromator can be used for wavelengths up to some 350 μm.

An interesting spectrometer which is becoming widely used in the i.r. is based on the Michelson interferometer. To understand how this instrument can operate as a spectrometer, consider a monochromatic source of radiation wavelength λ. Let a beam from the source be divided into two, a division normally made by a partially reflecting, partially transmitting element, one beam being subsequently reflected from a mirror to rejoin the other beam at a radiation detector. If the path lengths of the two beams are equal then they will reinforce each other at the detector but, if the mirror be moved so the path lengths differ by $\lambda/2$, then the two beams will interfere destructively. By moving the mirror at some fixed speed u, the reinforcement and destruction will take place regularly at a time-interval t where

$$ut = \lambda/2. \qquad (2.10)$$

This will give rise to an oscillating detector output with a frequency u/λ. If a second source of radiation, having a wavelength $N\lambda$ and an intensity M times that of the first source, is added, the output from the detector will contain a component with the frequency $u/N\lambda$ and M times the amplitude (Fig. 2.12).

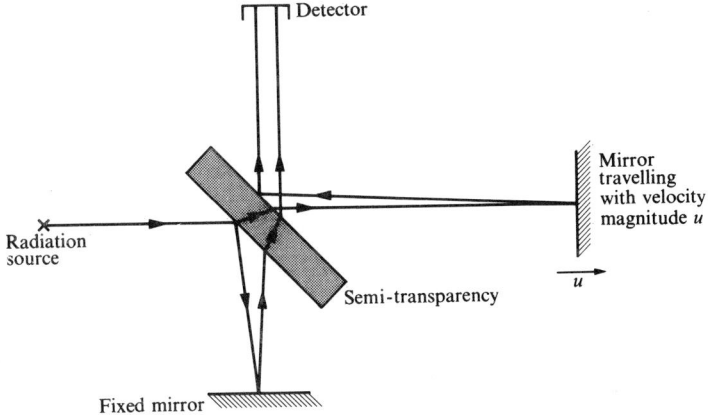

Fig. 2.12. Schematic of Fourier-transform spectrometer.

This concept can be extended to all the radiation wavelengths present in a real source, each contributing a signal of appropriate intensity and frequency. The total mixture must then be separated into its component parts by Fourier analysis, usually performed by computer, and it is the tedious nature of this analysis in the absence of suitable computers which delayed the widespread appearance of Fourier-transform spectroscopy until the 1950s, in spite of its concept in 1891 by Michelson.

To use such an instrument as a true monochromator is evidently impossible, but some aspects of monochromation can be simulated by restricting the response of the detector to a fixed frequency. Supposing, for example, it was required to observe only the effect of radiation having a wavelength λ on a system, then the detector would be electronically constrained to respond only to frequencies near u/λ. It must be remembered that the presence of other wavelength radiation in the beam can affect the system under investigation by altering the transparency of materials, or even detector response, and in this way it falls short of true monochromation.

2.6. Polarization in the infrared

Before considering the final act of radiation absorption at the detector, we will examine the means for polarizing the radiation if necessary. Defining the plane of polarization as that containing the electric vector then the reflection from a material at the Brewster angle can be used, as for visible radiation, to suppress the electric vector in one plane and, if such a reflection is repeated sufficiently the vector will eventually be confined to one plane, although not without loss of power at each reflection. The complement to this form of polarizer is one in which the transmitted, not the reflected, beam is used to give polarization at right angles to that in the reflected beam. For the latter 'transmission' polarizer polyethelene is a favoured material in view of the need for transparency, while for reflective polarizers such considerations are absent and any good metallic surface will serve.

For the whole of a radiation-beam to strike a polarizer at the Brewster angle it is necessary that complete parallelism should obtain in the beam, a requirement which may be experimentally difficult to satisfy. For longer wavelengths, say about 15 μm, sets of parallel metallic bars makes a more convenient and very efficient polarizer for which parallelism is less important. Bars about 5 μm wide separated by 5 μm spaces would be typical and these must be formed on i.r. transparent substrates such as thin polyethelene or proprietary 'Irtrans', as unsupported structures would be too fragile. Radiation polarized perpendicular to the axis of the bars is transmitted through the grating while that polarized parallel to the axis is reflected.

2.7. Absorption of infrared radiation on thermal detectors

To obtain the maximum benefit from all the processing the emitted radiation is assumed to have undergone it must now be completely absorbed by the detector. For the electronic type of detector, which will be described in the next chapter, radiation of the appropriate wavelength will be absorbed in consequence of the fundamental properties of the photosensitive material provided suitable antireflection films are present. The thermal detector must however offer a surface treated to absorb any radiation falling on it and this treatment usually involves the deposition of a film on the device. The surface film produced is loosely referred to as 'black', implying that it absorbs all i.r. radiation in the way that an optically black surface would absorb all visible radiation. The efficiency of such surfaces is due to their irregular, granular construction, which is encouraged in order that incident radiation be reflected to and fro in the surface area until it is wholly absorbed. The linear dimensions of the required irregularities may be comparable with the radiation wavelength and hence in the i.r. the surface may be composed of grains with sides of 10 μm or more. These grains are typically of gold, zinc, bismuth, or carbon, the importance of the material arising mainly from the manner in which it facilitates production of the required grain geometry. If radiation of one wavelength only is to be observed by a thermal detector it is possible to arrange a very efficient absorbing device using a film with an electrical resistivity of about 380 ohms per square set a distance equal to one quarter of the radiation wavelength before a reflecting plate. The complex of film and reflector forms a resonant space in which standing waves are set up and all the incident radiation is absorbed (Fig. 2.13). Such a receiver cavity working effec-

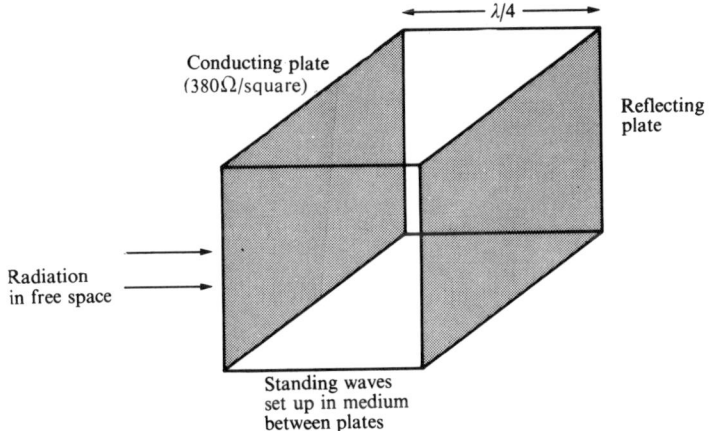

Fig. 2.13. Total absorber for radiation of wavelength λ.

tively at only one wavelength vitiates the special 'all-wavelength' detection which thermal devices normally show, a virtue discussed in detail in Chapter 4. To retain this while still absorbing at least 50 per cent of the incident radiation it is necessary to use a conducting film with a resistivity of about 180 ohms per square. The universality of this result is surprising but an insight is given by considering the fraction of radiation that would be transmitted by any metallic film. It would be very large, approaching unity for a thin film and falling to zero as the film thickness increased. Conversely the reflectivity of the thin film is small, increasing to a large, metallic, value when the film is very thick. The fraction of radiation absorbed is the difference between unity and the sum of the transmitted and reflected components. This absorbed fraction increases from zero in a very thin film, where all is transmitted, and falls again to zero in a thick film, where all is reflected. Between these extremes absorption reaches 50 per cent with a film thickness having the stated resistivity of ~180 ohms per square.

3 Vacuum and solid-state electronic detectors

3.1. Electronic photodetectors

The electronic photodetector relies for its operation on the liberation of a charged particle from some parent body by an incident photon. It will be found more convenient here to consider light as a particle stream and consider photon–electron interactions rather than electromagnetic-radiation–electron interactions. The liberation of the charge-carrier may be either total, removing it completely from its parent body, or partial, wherein a previously immobile charge-carrier gains freedom of movement within the body.

Devices in the first category (vacuum photocells) consist essentially of a metallic or semiconducting plate, the target or cathode, upon which falls the photon stream to be detected. The incident photons free electrons from the target surface, these are attracted to an adjacent, positively charged electrode, and current can then flow between the two (Fig. 3.1). In its simplest form this type

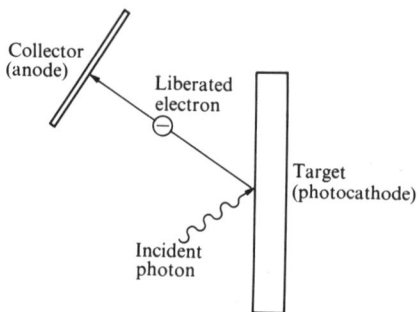

Fig. 3.1. Basic mechanism of vacuum photodetector.

of detector appears to be falling out of fashion, since it requires an evacuated space to allow unimpeded electron passage between the two electrodes, and the consequent fragile glass envelope needed to maintain the vacuum is an obvious disadvantage. However, one highly developed member of this category, the photomultiplier, flourishes by virtue of its great sensitivity.

In this instrument the freed electron is accelerated towards the anode so strongly that, on impact, a shower of secondary electrons is released, and these in turn are accelerated towards a further electrode, now rechristened a dynode, where

Vacuum and solid-state electronic detectors

the process is repeated. By doing this several times the first, single electron obtained can appear at the final electrode multiplied by as much as 10^6. The image-converter is another variant and in this a pattern or image of low-energy photons falling on the target gives rise to a corresponding pattern of free electrons, which are accelerated towards a screen where they excite visible fluorescence, thus reproducing the original i.r. image.

In the second category, where the electron is not fully freed but only mobilized within the solid body, we have a bewildering number of ways in which it can be utilized. Fortunately, only two or three are of widespread significance, but before discussing any of them it is necessary to appreciate that strange mathematical entity, the positive hole. The electrons in a semiconductor are only allowed to assume energies which lie in certain well-defined bands. The lower-energy bands are not of interest to us, but the highest, or conduction, band and the valence band lying immediately below it energetically are the operative regions in photodetection. At any finite temperature the conduction band will have a few electrons in it and these will be free to move under the influence of an electric field. Now these electrons have come from the valence band and their departure will have left gaps in the bulk of electrons which, at the absolute zero of temperature, would have filled this band (Fig. 3.2). It is these gaps which act like positively

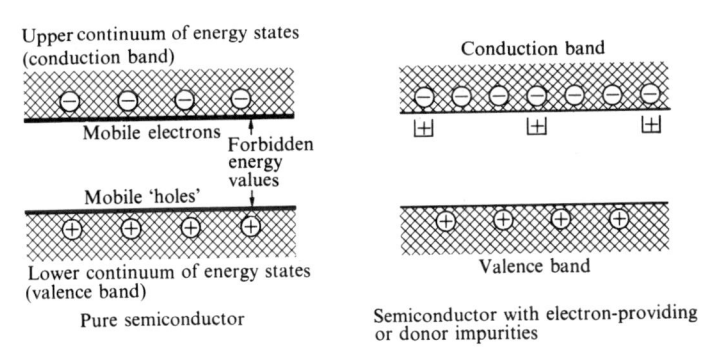

Fig. 3.2. Charge-carriers in a semiconductor.

charged carriers moving under a field in a direction opposite to that of the electrons. Analytic convenience is served by regarding these holes as the working entities rather than the remaining electrons, rather in the manner that the movement of a spirit-level bubble against gravity is more simply estimated than that of the surrounding liquid. It has been tacitly assumed that the electrons in the conduction band have been excited from the valence band by thermal energy but, in addition to these, others can be raised to this level by photons incident on the material and it is these which must be made manifest by some external effect for a useful photodetector to be made. One of the first observed and still the most widely used of external effects is the increased electrical conductivity which the material shows (the photoconductive detector). Nearly as common is the method whereby a strong electric field is 'built into' the semiconductor and photoexcited electrons which drift into it are driven round an external circuit (the photovoltaic detector). The only other technique to be examined in detail exploits the excitation of a high density of electrons and holes on one side of a semiconducting plate. These diffuse away in a strong transverse magnetic field which accelerates them round the external circuit in a direction normal to the plane containing the magnetic field and diffusion direction (the photo-electro-magnetic detector).

The most significant characteristics of a detector are the wavelength region to which it responds, the speed of that response, and the magnitude of the signal it provides. For the moment we shall ignore the noise, i.e. the apparent random signal present when the detector is not being irradiated, since this undesirable intruder is best dealt with in conjunction with the amplifying circuits used, for these will introduce their own added quota of noise.

3.2. The vacuum photocell

The vacuum photocell wavelength response is controlled by the amount of energy needed to release an electron from the target material. This energy is called the work function ϕ and the longest wavelength radiation to which the cell will respond is $1 \cdot 2/\phi$ μm if ϕ is measured in electronvolts. The work function of any particular material can be reduced, and hence the longest wavelength response increased, by treating the surface, the deposition of caesium being particularly effective for this purpose. Characteristics of a number of target materials are shown in Table 3.1. The factors which affect the speed of response of the cell

TABLE 3.1

Target material		Wavelength in μm by which quantum efficiency has fallen to 10^{-4}
Na$_2$KSB	Caesium treated	0·8
GaAs	Caesium treated	0·9
Ag	Caesium treated	1·0
In As P	Caesium treated	1·1

are the time taken for an incident photon to produce a free electron and then the time for this electron to traverse the space between the target and anode. The actual production takes only about 10^{-13} second and can be ignored, leaving the target—anode transit time as the dominating factor or, in the case of the photomultiplier, the sum of the dynode—dynode transit times.

It is in fact the variation in this time that is significant, since in practice we are concerned with obtaining a signal that accurately reproduces an incident radiation pulse. It is evident that if the time between photon impact and final electron collection is unchanging, then, however long it may be delayed, the pulse shape will be reproduced accurately.

Vacuum and solid-state electronic detectors

A comment applicable to all detectors must be entered here. There is a fundamental limit to the response speed arising from the electrical resistance R and capacity C that any real device will have. To understand this, first consider a simple series combination of resistor and capacitor as in Fig. 3.3(a).

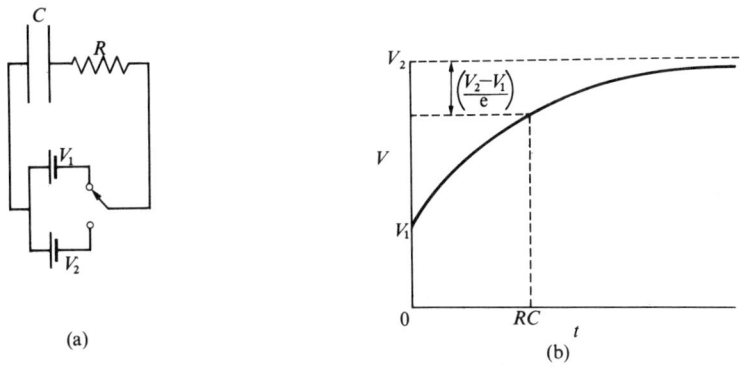

Fig. 3.3. RC response speed

If at zero time the potential difference held across the terminals is instantaneously increased from V_1 to V_2 the potential V across the capacitor will rise in obedience to the equation

$$V = V_1 + (V_2 - V_1)(1 - \exp(-t/RC)). \tag{3.1}$$

This is shown in Fig. 3.3(b).

The equation shows that V will not reach its equilibrium value of V_2 until an infinite time has elapsed, and so to obtain a finite figure which can be used to compare different resistance–capacitance combinations we calculate the time taken to get within $1/e$ of the total change of potential. This occurs when $t = RC$, it being noted that the dimension of resistance × capacitance is time. If R is measured in ohms and C in farads the units of t are seconds.

If this R–C combination is part of a complete circuit then that circuit will not be able to respond to any change in conditions in a time less than RC whatever other components are involved.

To return now to the particular case of a detector: it can now be seen that any change in the radiation falling on it cannot be made evident until a period of at least RC has elapsed.

This value is referred to as the RC limit and may override other considerations. Certainly in practice vacuum photocells are usually dominated by their RC limit, which is of the order of nanoseconds.

The signal from such a cell will depend on the quantum efficiency η of the target. This is defined as the reciprocal of the number of photons which must be incident in order for one electron to be liberated and is normally about 0·25, i.e. about four photons are needed for one successful liberation. If we have

F photons per second falling on the cell a current of $F\eta$ electrons per second will flow in the external circuit giving a signal current of I where

$$I = qF\eta \text{ ampere}, \qquad (3.2)$$

and q is the electronic charge ($1 \cdot 6 \times 10^{-19}$ coulomb). This current is passed through an external load resistance R_L and the potential difference developed across the load forms the observed signal. The magnitude of this signal voltage clearly increases with increasing resistance, but at the same time the response speed will fall, as the device is now assumed to be RC limited. This choice between signal magnitude and response time is a common dilemma for photodetector users and has no general solution. The very high signal that the photomultiplier affords, approximately 10^6 times that of a simple vacuum photocell, allows low load-resistances to be used without diminishing the signals to unrealistic values. Because of these low resistances, very rapid operation may be obtained with a spread of response time of as little as 10^{-14} second.

3.3. The semiconductor photodetector

(a) *General considerations*

Turning now to the solid-state detector, as opposed to the vacuum cell, the factor controlling the long wavelength sensitivity is the 'band-gap', by which is meant the energetic separation between valence and conduction bands. If an electron is to be excited from valence to conduction band by a photon then clearly the energy of the photon must equal that of the band-gap.

An important practical consequence of this necessarily small band-gap is the ease with which thermal energy can excite electrons across it. These would be indistinguishable from electrons that had originated with a photon and would thus obscure the true signal. It is because of this that electronic detectors require to work at low temperatures in the i.r., and a crude guide would suggest that the energy of the photon to be detected should, at least, be greater than the phonon energy in the semiconductor lattice, i.e. $1 \cdot 2/\lambda > kT$, k being the Boltzmann constant, T the absolute temperature, and λ the radiation wavelength in μm. (See Fig. 3.4.)

The elemental semiconductors silicon and germanium have band-gaps of about $1 \cdot 1$ and $0 \cdot 7$ eV respectively, which will put their longwave responses at $1 \cdot 24$ and $1 \cdot 8$ μm. This, however, is only brushing the extreme end of the i.r. spectrum and to reach further it is necessary to use semiconductors compounded of two or even three elements. While binary or two-element compounds have been very useful in reducing band-gaps down to say $0 \cdot 2$ eV, or about 6 μm response, a great flexibility is obtained when ternary or three-element compounds are used. This arises because certain ternary compounds may be made by mixing, in any proportion, two binary compounds having a common element. For example, consider cadmium telluride CdTe, and mercury telluride, HgTe. The first has a

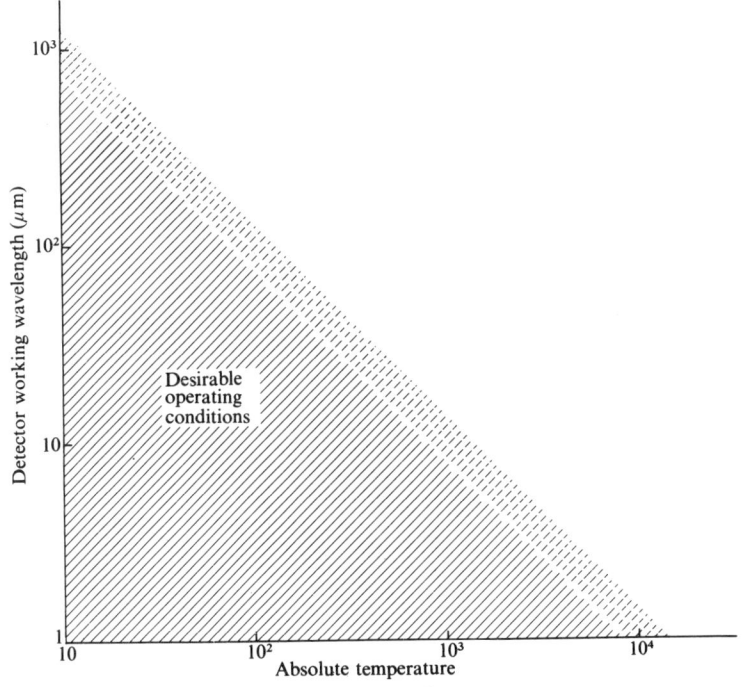

Fig. 3.4. Detector operating temperatures

band-gap of about 1·5 eV, while HgTe has a practically zero gap. As the compounds form a solid solution, appropriate admixtures of the two can theoretically be made to give band-gaps ranging from 1·5 eV to zero. A particularly interesting and efficient system is derived from lead telluride, PbTe, and tin telluride, SnTe. Here, although both materials have band-gaps of ~ 0.3 eV, alloys of ~ 40 per cent SnTe with 70 per cent PbTe show an almost zero gap. The magnitude of the band-gap is seen, in fact, to fall towards zero as the SnTe content of the ternary increases from nil, and then starts to increase again when concentrations of SnTe greater than about 40 per cent are used, (Fig. 3.5). A list of semiconductor alloys and elements useful for i.r. photodetectors would make dull reading and so further materials are relegated to Table 3.2 in company with those already mentioned.

Some digressions into simple semiconductor theory are inevitable in a book dealing with the i.r. and all that can be done is to make them as terse and lucid as possible. Such an excursion is necessary to explain the great change in cut-off wavelength shown by pure and impure germanium in Table 3.2. It is now necessary to admit that the band-gap, that region between conduction- and valence-band energies where no electron could go, is violable. If impurities in small quantities, say 1 in 10^4, are introduced into a semiconductor they may form permitted energy levels for electrons or holes in the gap and, significantly, these could be very close

Vacuum and solid-state electronic detectors

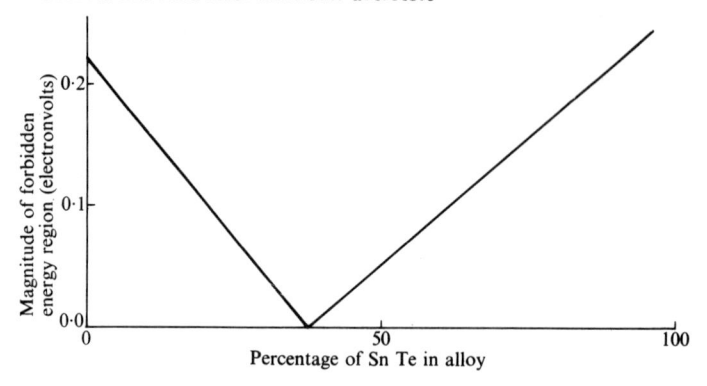

Fig. 3.5. Relationship between composition of PbTe–SnTe alloy and forbidden energy at 77K.

TABLE 3.2

Material	Working temperature (K)	Approximate cut-off wavelength (μm)
Ge	300	1·8
Si	300	1·2
Ge + Cu impurity	18	30
Ge + Au impurity	35	15
InAs	77	3·0
InSb	77	5·5
Cd–Hg Te	77	13
Cd–Hg Te	20	40
Pb–Sn Te	77	13
Pb–Sn Te	12	17

to the bottom of the conduction, or to the top of the valence band (Fig. 3.2). Although carriers can reside in these levels, some penalty for breaking the rules must be paid: they cannot move, and hence cannot conduct electricity, unlike the mobile populations in conduction and valence band.

Despite this they can become effective by photon excitation into a region of mobility in a manner similar to that in which the native electrons were considered to be excited from valence to conduction band. It can now be seen that if the copper or gold impurities in the germanium form electron-holding centres (known as donors) very close to the bottom of the conduction band then these could be excited by photons of a much lower energy than those needed for valence to conduction band stimulation.

(b) Photoconductive cells

The first intimations of the photoconductive effect were received by Willoughby Smith in 1873, when he observed that a standard selenium resistance, when used in sunlight, had a value which was less than that shown when used in shade. The

Vacuum and solid-state electronic detectors 29

semiconductor selenium still functions in photocells but, unfortunately for the logical development of this chapter, mostly in photovoltaic rather than photoconductive devices. We have seen how simple the operation of a basic photoconductive cell is, with a change of resistance due to the mobilization of charge-carriers by incident photons. To look a little more closely at the signal and response time, consider a basic series circuit of battery, current-meter, and photosensitive material (Fig. 3.6). A hole and electron pair that have been photon

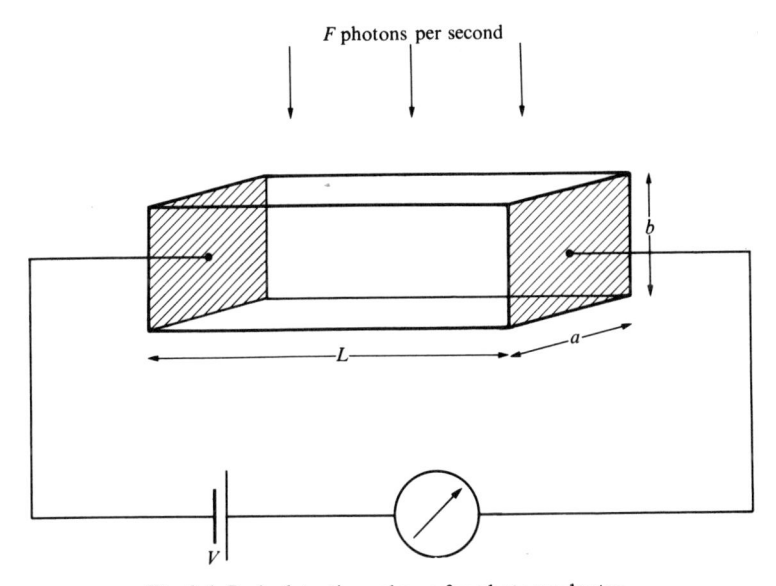

Fig. 3.6. Basic detection scheme for photoconductor.

excited into mobility will allow a current to flow in the circuit for the period which elapses before they recombine with one another to yield the fixed valence-band electron which was their starting point. To get a larger signal current it is necessary only to make the applied voltage from the battery larger, but there is unfortunately a limit to the useful increase that can be brought about in this way, and it arises from the nature of the contacts at the ends of the photodetector. If these are designed to allow the free passage of electrons then they will offer an obstruction to the holes which will be attempting to flow in the direction opposite to that of the electrons. Conversely, if the contacts allow hole passage then electrons are impeded. To keep the syntax reasonably manageable let us consider contacts which pass, or are ohmic to, electrons. Then holes will crowd up round one contact and the electrons reaching it have a greatly enhanced probability of recombination which destroys them. Thus it is seen that a voltage just great enough to cause a mobilized electron to traverse the detector in its lifetime, is the greatest that can be usefully applied. From the

30   Vacuum and solid-state electronic detectors

diagram, a voltage V will give a field V/L and a carrier velocity $\mu V/L$, where μ is the mobility of the carrier, or its velocity in unit field.

Time for traversal of detector $= L^2/\mu V$.

Since this must not be less than the lifetime of the hole–electron pair then the highest useful voltage is $L^2/\mu \tau$. If F photons per second fall on and are absorbed by the face aL they will generate carrier pairs at the rate of ηF. In equilibrium this generation rate must equal the recombination rate, i.e. the carrier population N divided by the carrier lifetime τ and so

$$\eta F = N/\tau \tag{3.3}$$

or

$$N = \eta F \tau \tag{3.4}$$

and the population density n is given by

$$n = N/abL = \eta F \tau / abL, \tag{3.5}$$

where a, b, L are dimensions shown in Fig. 3.6.

The conductivity of the specimen will thus be $\mu(\eta F \tau / abL)q$ and with the highest useful voltage $L^2 \mu / \tau$ applied, a signal current I_S is obtained where $I_S = \eta F q$. With the usual RC limitation warning, the response time is evidently equal to τ, the lifetime of a hole–electron pair. Summarizing then, a flux of F photons per second gives a signal current of $\eta F q$ amperes with a response time of τ seconds.

It is possible to remove the applied voltage limitation considered above by adding impurities to the semiconductors which yield 'traps' in the forbidden energy region. These traps must have the property of capturing a hole (or an electron) and holding it immobilized for a substantial time. While the hole (or electron) is kept out of play the remaining mobile electron (or hole) can traverse the photodetector as often as desired without risking premature recombination near the electrodes. It looks now as if we have no limit to the signal which can be made available by increasing the applied voltage but, perhaps not surprisingly, an infinite signal is not in fact obtainable. If very high fields are applied to a conducting medium then charge may be drawn into it too rapidly for it to maintain uniform density and a region of high charge density builds up near one electrode. This effect begins to be appreciable when the relaxation time, τ_R of the conducting medium approaches the carrier transit time $L^2/V\mu$. Since τ_R is the ratio of dielectric constant ϵ and electrical conductivity σ ($\tau_R = \epsilon/\sigma$), and we have already found σ to be $\mu(F\tau/abL)$, where τ now has the increased value due to the traps, then the maximum voltage that can be applied before space-charge limitation sets in is given by:

relaxation time of medium = carrier transit time

$$\epsilon/\sigma = L^2/V\mu \tag{3.6}$$

$$\epsilon/\{\mu(\eta F \tau/abL)q\} = L^2/V\mu \tag{3.7}$$

or
$$V = Lq\,(\eta F\tau)/eab. \tag{3.8}$$

The signal current I_S is now obtained as before and
$$I_S = (q\eta F\tau)^2 \mu/eab. \tag{3.9}$$

Although the introduction of traps has increased the maximum signal current available it has been done at the expense of making the effective lifetime τ greater, that is to say, by making the device slower. An ingenious method of evading this drawback known as the microwave-bias technique has now been evolved, but as this is atypical of photoconductive systems in general, it will be dealt with later in a chapter devoted to the non-classifiable.

(c) Photovoltaic cells

Turning now to the photovoltaic detector it is necessary to consider how the built-in voltage mentioned earlier can be effected. Suppose a specimen of semiconductor has been populated with impurities that yield mobile electrons (or doped n-type as the trade has it), and this is joined to a sample containing hole-yielding impurities (doped p-type). The excess electrons will diffuse into the hole-bearing material and conversely the excess holes will diffuse into the electron-bearing sample and the two parts will take up, respectively, a positive and negative charge, giving a small region of high electric field of the order of 10^5-10^6 Vm^{-1} at the interface. The techniques involved in making these junctions are the sophisticated core of the semiconductor industry, but our business lies with their use rather than their fabrication. In the photovoltaic detector the photon flux is allowed to fall on the surface of the junction-bearing material and the excited carriers diffuse to the nearby junction, where they enter the high-field region and are accelerated across the junction. If each arm of the junction is connected to an external circuit a current will flow. Provided the external circuit has a resistance considerably smaller than that of the junction then most of the excited-carrier flux will appear in it and the signal current will be simply given by
$$I_S = q\eta F. \tag{3.10}$$

This assumes that the junction is close enough to the irradiated surface for all the excited carriers to diffuse to the high-field region before they are diminished by recombination. In practice, the construction of the device usually allows this condition to be approached.

It may be more convenient electronically to use an external circuit of high impedance and amplify the voltage that is developed across it (the open-circuit mode). Let the external resistance be R_A and the detector resistance R_D: then the current flowing through R_A will be $\{R_D/(R_A + R_D)\} \times q\eta F$ and hence the signal voltage V_S appearing in the external circuit will be given by
$$V_S = \{R_D/(R_A + R_D)\} \times q\eta F \times R_A \tag{3.11}$$
$$R_D q\eta F \text{ if } R_A \gg R_D.$$

Vacuum and solid-state electronic detectors

The response time of this device will be the time taken for an excited carrier to diffuse into the junction region from the surface at which it originated. The message of any disturbance in the carrier population can be said to travel a distance X in the semiconductor with an effective velocity D/X, where D is the coefficient of diffusion for the carrier. So, if the junction and the irradiated surface are separated by a distance L, the response time τ will be given by

$$\tau = L^2/D. \tag{3.12}$$

The magnitude of D is related to the carrier mobility by the Einstein equation $D = kT\mu/q$ and so

$$\tau = L^2 q/kT\mu. \tag{3.13}$$

It is clear that if this time approaches the carrier lifetime then the signal will disappear since all the carriers will have recombined before reaching the high-field region, therefore in this case, carrier lifetime is not a limiting factor in response speed. To set against this, the inevitable RC limit is very significant in a photovoltaic detector. This is because the high-field region is denuded of charge-carriers by the field, and behaves as a rather poor dielectric slab which will appear as a capacitor in the circuit. Thus the device has both substantial resistance and capacitance, unlike the photoconductor, which is practically pure resistance.

The avalanche photodiode bears a relationship to the simple photovoltaic detector similar to that existing between vacuum photocell and photomultiplier. If an external voltage is applied to a photovoltaic device the built-in high-field region can be extended into the two arms of the junction. In this way the width of the region may be increased by a factor as great as 10^2 and the photoexcited carriers, accelerated by the high electric field, will have sufficient energy to collide with the semiconductor lattice and free even more carriers. This multiplication of the carriers is not generally allowed to exceed a value of $\sim 10\text{--}100$ times because of the onset of significant electrical noise, i.e. random currents not directly related to the signal. If the multiplication factor is m then, in a manner similar to that used to derive the photovoltaic signal current, we obtain

$$I_S = q\eta F m. \tag{3.14}$$

The wide high-field region, which it will be recalled is without free charge-carriers, gives the avalanche photodiode a high resistance and because of this it is not normally used in the open circuit, or voltage, mode. A very useful consequence of this wide, high-field region is the low capacity that it gives to the device. This, in practice, outweighs the effect of the larger resistance and the response is limited only by the time taken for a carrier to drift across the high-field region. If this field is very high then the drift velocity will cease to be proportional to the applied field and saturate at some value V_{sat}. This saturation velocity is usually about 10^5 ms^{-1}, close to the value of the carriers' thermal velocity. The time for crossing a region W is τ, where

$$\tau = W/V_{\text{sat}}. \qquad (3.15)$$

A value of 10^{-10} is typical for τ. It is clear that the avalanche detector is very fast and sensitive but the actual fabrication requires extremely skilled technology. This is mainly because the high voltages involved can only be applied to pure, homogeneous material without provoking localized breakdown, and, further, device geometry must be such that exposed surfaces have little field developed across them. Because of these difficulties, avalanche detectors are at present only manufactured in silicon, with its 1·2 μm wavelength limit, although germanium and gallium arsenide have been used in laboratory experiments.

(d) Photo-electro-magnetic cells

The photo-electro-magnetic detector (P.E.M.) is unusual, although not unique, among the devices we are considering in requiring an external magnetic field for its operation (Fig. 3.7). As usual we will suppose a photon flux F with quantum

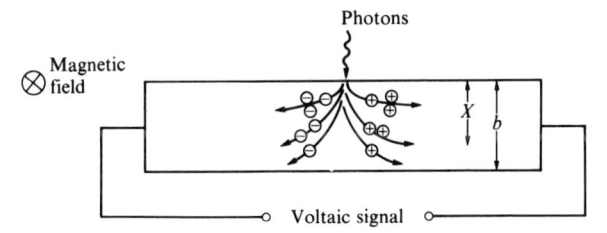

Fig. 3.7. Photo-electro-magnetic detector.

efficiency η to fall on one face of the detector. The current which will diffuse away from this face will be $2\eta Fq$ amperes carried by both holes and electrons. This current will decay due to recombination of the excited carriers as it moves into the device and, at some distance X away from the irradiated face it will have a value $2\eta Fq \exp(-X/L^*)$. L^* is called the ambipolar diffusion length and is defined as the distance that the population excess created at the front face can diffuse before it has decayed to 1/e of its value. The mean front to back current flow is then:

$$\frac{1}{b}\int_0^b 2\eta Fq \exp(-X/L^*) dX = \{2\eta Fq\, L^*/b\}\{1 - \exp(-b/L^*)\}. \qquad (3.16)$$

The magnetic induction B deflects this current through an angle μB towards the sides of the device giving a lateral current

$$\{2\eta Fq\, L^*/b\}\{1 - \exp(-b/L^*)\} \tan \mu B.$$

Here μ is the mobility of the hole–electron cloud and μB is generally small enough for the tangent approximation to be used. In practice b is made $\sim L^*$ and the exponential term is about ½.

The signal current I_S can now be written

$$I_S \sim \frac{\eta F q L^*}{b} \mu B, \qquad (3.17)$$

and signal voltage V_S

$$V_S \sim \frac{\eta F q L^*}{b} \mu B R_D, \qquad (3.18)$$

where R_D is the device resistance.

The ambipolar diffusion length L^* increases with increasing temperatures, as most diffusion lengths do, and because of this it is evident that the P.E.M. device may, with advantage, be operated at temperatures higher than those used for photovoltaic or photoconductive cells. Variation in wavelength response and in the density of thermally excited carriers must, of course, be remembered when selecting an operating temperature.

The response time of the P.E.M. detector is best considered by examining the situation when the light signal falling on it is stopped. A gradient of excited carriers, dense at the front and less dense to the rear of the device, will exist, and while this gradient remains the signal current will flow but the population will, by diffusion, tend towards uniform density. Using arguments similar to those applied to the photovoltaic response time we see the message of non-uniformity travels from front to back of the device in a time b^2/D, or, replacing the diffusion coefficient D by $kT\mu/q$, a time $b^2q/kT\mu$. For such a period, then, the current will be maintained unless the excited carriers have first been destroyed by their normal recombination processes. Hence the response time τ_D is either τ the carrier lifetime or $b^2q/kT\mu$, whichever is the shorter.

$$\tau_D = \frac{b^2 q}{kT\mu} \text{ or } \tau.$$

While discussing the behaviour of these devices at least two implicit assumptions have been made in order that the mechanisms involved were not obscured by superfluous detail. We will now examine the assumptions, their degree of validity, and the qualitative effects they have had on the analysis.

Firstly, all the incident radiation usefully absorbed was assumed to be converted into free carriers at the front face of the detector. In practice the radiation will penetrate the device with an exponential decrease in intensity and carriers will be excited throughout the material. However, working with wavelengths at, or shorter than, the detector limit will mean that the radiation is absorbed very heavily in a short distance and our approximations are good. If longer-wavelength radiation is used, it will pass through the device without interacting in any way to produce photoexcited carriers and so this situation is not of interest. The overall effect of this first assumption is not then significant as far as signal is concerned

Vacuum and solid-state electronic detectors

but will give a slightly longer response time in P.E.M. detectors than practice would show, unless they are carrier-lifetime limited anyway.

Secondly, the special nature of the semiconductor surface has been ignored and it has been regarded as behaving exactly as the bulk. In fact the free surface will generally show a very short carrier lifetime, or fast recombination rate, and the photoexcited carriers, being close to a surface, will recombine with greater rapidity than carriers in the bulk. Although elaborate polishing and passivating techniques are applied to detector surfaces to minimize this the signal values obtained above will be a little optimistic.

A rigorous consideration of these factors, together with the influence of unsolicited carrier traps, will enable expressions of a magnificent complexity to be evolved for signal and response. Interested readers will find such rigour in some of the volumes listed at the end of the book for further reading.

Most electronic detector types in use will be found among the devices dealt with in this chapter and their signal and response have been collected in Table 3.3 for ready access and comparison.

TABLE 3.3
For detector with F photons per second incident

Device	Current signal	Voltage signal	Response time
Photoemissive	$\eta F q$	$\eta F q R_L$	Usually RC limited
Photomultiplier	$\eta F q m$	$\eta F q m R_L$	Usually RC limited
Photoconductive (no impurity)	$\eta F q$	$\eta F q R_L$	τ carrier
Photoconductive (activating impurity)	$\dfrac{(\eta F q \tau)^2 \mu}{\epsilon a b}$	$\dfrac{(\eta F q \tau)^2 R_L}{\epsilon a b}$	τ carrier (enhanced)
Photovoltaic	$\eta F q$	$R_D \eta F q$	$L^2 q/kT\mu$ or τ (carrier) if smaller
Avalanche	$\eta F q m$		W/V_S
P.E.M.	$\eta F q L^* \mu B / b$	$\eta F q L^* \mu B R_D / b$	$b^2 q/kT\mu$ or τ (carrier) if smaller

4   Thermal detectors

4.1. Detection by thermal effects

Thermal detectors, as their name implies, depend on the heating effects of incident radiation; when considering them we shall ignore the photon aspect and deal only in terms of the power in the incident i.r. signal. This fundamental difference in operation gives them particular virtues and vices in the major characteristics of sensitivity, wavelength response, and speed. The sensitivity will generally be lower than that of an electronic detector but, to offset this, the detector may often be worked at room temperature without refrigeration. The wavelength response is normally much broader and displays a less precise cut-off since, provided the signal radiation can be absorbed in the device, the magnitude of actual energy quanta in it is irrelevant. The response speed is smaller, as might be expected, for it will be, in effect, dependent on the time taken to heat a finite piece of material, whereas it will be recalled that the photon—electron energy exchange requires only some 10^{-13} second. Skilled fabrication can minimize this thermal time constant, but seldom to a point where it can compete with that of the electronic device.

Three types of thermal effect are used and the first, employed by the largest group of thermal detectors, is the absolute rise in temperature brought about by incident radiation. These detectors are in fact sensitive thermometers. The Golay cell, a sealed gas container, responds to the temperature increase of the gas under irradiation by displacement of a small flexible diaphragm forming part of the container. This displacement is sensed and related to the incident power. Probably the oldest member of this group, the bolometer uses variation of electrical resistance with temperature to give an instrument which, from time to time, enjoys a new lease of life when the material technologist produces a substance with a yet greater temperature coefficient of resistance. The thermocouple, used either as a single junction or as a series of connected couples, is frequently employed, especially to measure, as opposed to merely detect, the power in an i.r. beam. Where very intense powers are involved, as in the monitoring of a laser output, a direct calorimetric method may be used and the rise in temperature of a small copper slug or carbon cone on which the radiation falls, is measured with a conventional thermometer.

Another type of detector relying on an absolute temperature change is the 'thermactor'. It will be remembered that where different types of semiconducting material (p-type and n-type) form a junction there is a region having very few free charge-carriers. This region, acting as a dielectric slab, gives electrical capacitance to the device and, if the formation of the junction satisfies certain rather stringent conditions, this capacitance will change rapidly with temperature. As electrical capacitance can be measured very accurately the effect has been used to make a sensitive detector. However, its use does not appear to have become very widespread, possibly because of difficulty in making the specialized junction in a predictable manner. While it is not intended to enlarge on the mechanism or performance of this device, as we shall later do with others mentioned, it should be noted that one is reported as having nearly reached the theoretical limit of sensitivity.

The rate of change of temperature is the second thermal effect to be considered, and is necessary for the exploitation of the pyroelectric effect. Certain crystals develop a surface electrical charge when their temperature is changed. This charge appears in equal and opposite amounts on crystal faces which are perpendicular to the pyroelectric axis. The dipole produced in this way by some finite temperature change would, if left, leak away or be neutralized by attracting small, oppositely charged, particles from its surroundings. If the crystal faces are connected together electrically however, the charge will flow through the connection, and this burst of current will be repeated each time the temperature is changed. If no external connection is made then the presence of the charge will be manifest as a voltage existing between the two crystal faces. This voltage is a consequence of the electric field that the charge separation in the crystal has caused, and it can be estimated quite simply by regarding the crystal as an electrical capacitor (most pyroelectrics are substantially dielectrics or insulators) and the voltage can then be estimated from the known charge. Even if this open-circuit voltage is used for the signal, as opposed to the closed-circuit current, it is still necessary to vary the temperature periodically as the charge would eventually leak away. Many pyroelectric materials have been fabricated and their characteristics investigated but the dominant compound is still triglycine sulphate (T.G.S.). This pre-eminence has been maintained because the factors affecting material performance are not only the pyroelectric coefficient itself, which, of course, governs the charge produced for a given temperature change, but also various thermal characteristics and it is the overall combination of these (in a form we shall later derive) which appears difficult to surpass.

The last sub-section of thermal use involves the consequences of setting up a spatial temperature gradient in a material. This method promises an advantage in response speed as a temperature gradient can be set up very rapidly compared with the time needed to raise a bulk of material through any finite temperature range. Unfortunately, the device that exploits this has not proved particularly sensitive and we will discuss it here without later analysis. In brief, a temperature gradient in a semiconducting material will cause a diffusive flow of charge-carriers in the direction of decreasing temperature. If this flow is subjected to a magnetic

38 Thermal detectors

field it can be deflected and measured as an open-circuit voltage in a manner analagous to that used in the P.E.M. detector. This 'Nernst Effect' detector has been fabricated in bismuth, having a small antimony impurity, and in nickel and cadmium arsenides. Response times of 10^{-7} second have been obtained.

4.2. The thermal equation

Before looking in any detail at detectors of the first two groups, i.e. absolute temperature change and rate of temperature change sensors, it will be convenient to perform a simple thermal analysis which can subsequently be quoted without interrupting consideration of matters specific to some particular device. Consider then a rectangular slice of substance area A, thickness h and volume specific heat s irradiated with a power $\psi \cos 2\pi ft$ (Fig. 4.1). If the slice is connected to an

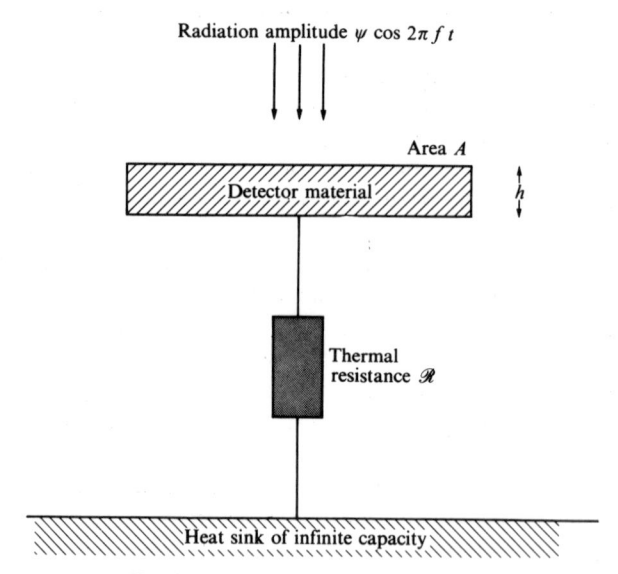

Fig. 4.1. Thermal circuit for detector.

infinite heat sink through a thermal resistance $\mathcal{R}$ then the heat balance equation is

$$\psi \cos 2\pi ft - \Delta T/\mathcal{R} = d/dt(\Delta T)\, Ahs. \qquad (4.1)$$

Here ΔT is the temperature of the slice measured with respect to the infinite heat sink and is assumed to have the same value throughout at any given temperature. The periodic solution of this heat balance equation is

$$\Delta T = \psi \cos(2\pi ft - \gamma)/\{(1/\mathcal{R}^2) + (hAs2\pi f)^2\}^{\tfrac{1}{2}}. \qquad (4.2)$$

The phase shift γ is equal to arc tan $(\mathcal{R}hAs2f)$ but is not of major interest. The general behaviour of ΔT might well have been deduced intuitively. It varies with

Thermal detectors

the same frequency, although not the same phase, as that of the applied radiative power. If the resistance of the heat sink $\mathcal{R}$ is very great then the slice can be regarded as thermally isolated and the excursions of T become very large.

$$\Delta T \to \psi \cos(2\pi ft - \gamma)/hAs2\pi f.$$

If $\mathcal{R}$ is very small, i.e. the specimen is well linked to the heat sink, then the excursions of T become small.

$$\Delta T \to \mathcal{R}\psi \cos(2\pi ft - \psi)$$

We can now discuss a number of thermal detectors in detail.

4.3. The Golay detector

The successful form of pneumatic detector made public in 1947 by M. Golay consists essentially of a gas container with a radiation absorber and a flexible diaphragm (Fig. 4.2). As radiation is absorbed the gas temperature rises and

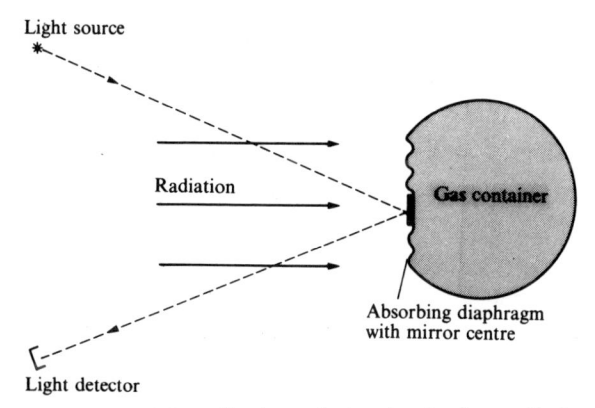

Fig. 4.2. Schematic of Golay cell using optical system to observe diaphragm.

the resultant increase in pressure causes deformation of the diaphragm. This diaphragm is silvered to form a mirror and any displacement can be observed optically, with high sensitivity. Let the detector be represented schematically by a volume of gas V at a pressure P and temperature T in a container having a diaphragm of area A. The restoring force on the diaphragm is such that a displacement x requires a deforming force αx. When the system absorbs some heat energy then $V \to (V + \Delta V)$, $P \to (P + \Delta P)$, $T \to (T + \Delta T)$, and the diaphragm moves a distance x.

For an ideal gas PV/T is constant and so

$$\Delta P = P\left(\frac{\Delta T}{T} - \frac{\Delta V}{V}\right). \tag{4.3}$$

40 Thermal detectors

This excess pressure must be consistent with the thrust on the displaced diaphragm

$$A\Delta P = \alpha x \tag{4.4}$$

and

$$\Delta V = Ax. \tag{4.5}$$

ΔP can be eliminated from these equations to give a value for x:

$$x = AP\Delta T/T(\alpha + A^2 P/V). \tag{4.6}$$

It is now necessary to relate the heat energy absorbed by the gas to the change in T. The thermal capacity of the gas is PV/T and this must replace the value hAS used in the introductory thermal analysis. The equation for a loosely linked system is chosen so that

$$\Delta T = \psi \cos(2\pi ft - \gamma) / \left(\frac{2\pi ft \, PV}{T}\right), \tag{4.7}$$

$$x = A\psi \cos(2\pi ft - \gamma) / 2\pi f (\alpha V + A^2 P). \tag{4.8}$$

The diaphragm is arranged to reflect light onto a photocell, the light intensity, and hence the photocell current I_S, being proportional to the displacement x, say Kx. Thus the signal current I_S is given:

$$I_S = KA\psi \cos(2\pi ft - \gamma) / 2\pi f (\alpha V + A^2 P). \tag{4.9}$$

This expression does not give any guide to the sensitivity of the detector since K would appear to be at the user's disposal by adopting appropriate optical systems and light intensities.

However, random motion of the gas molecules will give a diaphragm fluctuation Δx such that the mean stored energy, $\frac{1}{2}\alpha\Delta x^2$ equals kT, the thermal energy,

$$\text{i.e. } \Delta x = \sqrt{(2kT/\alpha)}. \tag{4.10}$$

From the equation for x it is seen that a power equal to

$$2\pi f (\alpha V/A + AP) \sqrt{(2kT/\alpha)}$$

would give this diaphragm displacement and, clearly, this is the smallest power which can be unambiguously detected.

$$\text{Minimum detectable power} = 2\pi f(\alpha V/A + AP)(2kT/\alpha)^{1/2}. \tag{4.11}$$

In practice this can be about 10^{-9} W.

Generally the gas is not held in a sealed container since this would cause the equilibrium position of the diaphragm, and hence the detector performance, to vary with the atmospheric pressure. A small leak to the atmosphere has therefore to be introduced and this prevents the value of $2\pi f$ being too greatly reduced since a long, sustained irradiation would only result in blowing gas out of the container. The lowest frequency that can be used in practice is in the order of 5 Hz. At the other end of the response scale frequencies of several thousand Hz can be used.

The wavelength response, as in all thermal detectors, depends on the ability of the gas container to absorb the incident radiation and it extends at least to several hundred microns. A very interesting variant of the Golay cell involves the replacement of absorbing surfaces by transparent windows allowing the radiation to penetrate to the gas itself. If the gas chosen for use has a strong absorption at one particular wavelength then the cell will show greatly enhanced sensitivity at that wavelength and little response at others. Such a selective device has been used to study detail in ammonia and carbon dioxide absorption spectra.

4.4. The bolometer

The prime requirements of the bolometer are a low thermal capacity and a high temperature coefficient of electrical resistance. Since 1880 when Langley first used blackened platinum strips much ingenuity has gone into satisfying the first consideration and contemporary instruments generally use small flakes of conductor only tens of microns thick. The material used may have a large metallic temperature coefficient of resistance, such as the platinum already mentioned, where resistance will increase rapidly with temperature or, conversely, the 'thermistor' uses a commercially unspecified mixture of semiconducting oxides to give an exponential decrease of resistance with temperature. A slightly different principle is involved when a discontinuity in resistance, associated with a change of state in the material, is employed. This change can be from a normal to a superconducting state as, for example, that displayed by columbium nitride at about 14·5K, this alloy showing a sixfold increase in resistance between 14·34K and 14·37K — a notional temperature coefficient of nearly $200K^{-1}$. The obvious penalty for this sensitivity is the need to maintain the element, very precisely, at the lower working temperature. A more convenient temperature region, about 350K, is associated with the transition from cubic to tetragonal crystal lattice of proprietary alloys made by Transco. This gives a similar order of temperature coefficient but the constant temperature necessary is more simply maintained, being above room temperature.

In use, two elements are employed, one exposed to the radiation and one shaded to act as a reference resistance and hence compensate for general temperature variations in the apparatus (Fig. 4.3). Some current must be passed through the element in order to observe the electrical resistance and its variation, and the magnitude of this current may be critical in as much as it forms a source of heat. If the current is maintained constant in an element having a positive temperature coefficient, then the heating effect, (current)^2 × resistance, will tend to increase even further. Similarly, if the voltage across an element having a negative temperature coefficient is maintained constant, the heating effect will

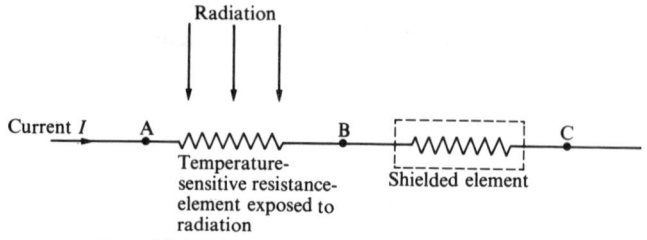

Thermal detectors

```
              Radiation
               ↓ ↓ ↓
Current I    A         B              C
   ●────────WWWWW──────●┌ ─WWWW─ ┐────●
           Temperature-   │              │
           sensitive resistance-  Shielded element
           element exposed to
           radiation
      Potential A–B is compared with potential B–C
```

Fig. 4.3. Essence of bolometer system.

tend to be unstable, the decreasing resistance increasing the Joule Heating, (voltage)²/resistance. The extent of these thermal 'run-aways' is, of course, dependent on the magnitude of the heat link $\mathcal{R}$.

To examine the signal from a bolometer we first estimate the change in resistance which our standard power flux $\psi \cos 2\pi ft$ would produce. The temperature coefficient of resistance β is related to the change ΔR in R, when the temperature T varies by ΔT, by

$$\beta = (1/R)(\Delta R/\Delta T). \tag{4.12}$$

or

$$\Delta R = R\beta \Delta T. \tag{4.13}$$

Using the general expression for ΔT then gives

$$\Delta R = R\beta \psi \cos(2\pi ft - \gamma)/\{(1/\mathcal{R}^2) + (hAs2\pi f)^2\}^{1/2}. \tag{4.14}$$

If a current I is passed through the bolometer the voltage signal across it is V_S where

$$V_S = IR\beta \psi \cos(2\pi ft - \gamma)/\{(1/\mathcal{R}^2) + (hAs\ 2\pi f)^2\}^{1/2}. \tag{4.15}$$

Again we have analysed the signal for cosinusoidally varying radiation. Although, unlike the case of the Golay cell, there is no overriding need for this variation it is normally more convenient to deal with an alternating electrical signal. This will be accorded more profound consideration in the chapter dealing with signal processing and noise. The apparently unlimited increase in signal is, as usual, formally available by increasing I. Practical limitations are imposed by the heating effect of this current and, in the last resort the small fluctuations in equilibrium temperature, which any body must display, need to be appreciably exceeded by the signal heating. Again this aspect is more closely scrutinized in the noise and signal processing chapter.

In practice powers of less than 10^{-10} watts can be detected when the value of $2\pi f$ is about 10^3 s^{-1}. This need to quote the speed of operation of the cell when stating its sensitivity is evident from a consideration of the equation for V_S. At low frequencies, i.e. provided $hAs2\pi f < 1/\mathcal{R}$ the signal amplitude is independent of f but when this inequality is no longer satisfied then V_S will approach a $1/f$ dependence.

The wavelength region which can be covered depends, as for all thermal detectors, on the absorption that is effected by the sensitive element and an evaporated metal film is available which will absorb radiation with a wavelength of up to several hundred microns.

4.5. Thermojunction and thermopile

The mechanism used to sense temperature rise in the thermopile is the thermovoltaic effect. Briefly, if two dissimilar electrical conductors are joined in a continuous loop, then an e.m.f. will appear in the circuit if the two junctions are at different temperatures. In this way, by keeping one junction at a fixed, reference, temperature and heating the other by irradiation, a signal e.m.f. can be made to appear in the loop. This technique was used by Melloni in 1835, some thirteen years after the thermovoltaic effect had been reported by Seebeck. Melloni, in fact, employed a chain of such 'thermocouples' arranged to expose alternate junctions to the radiation while the remaining junctions were shielded. This method of multiplying the signal voltage by using a pile of thermocouples, or thermopile, is usually adopted now.

The factors of importance in thermopile design are the thermal capacity of the device which, as usual, must be kept low; the magnitude of the voltage generated in the circuit for one degree temperature difference between the junctions (known as the thermoelectric coefficient) which must be large; the thermal conductivity of the junction materials which should be low to maximize the temperature difference which can be maintained between the junctions and, lastly, the electrical conductivity which should be high to maximize the power available from the device.

Examining these factors formally we have for the signal voltage V_S,

$$V_S = \mathscr{S} \Delta T \qquad (4.16)$$

where $\mathscr{S}$ is the Seebeck or thermoelectric coefficient. Putting the usual value for ΔT in this equation gives

$$V_S = \mathscr{S} \psi \cos(2\pi ft - \gamma)/\{(1/\mathscr{R}^2) + (2\pi f \, kAs)^2\}^{1/2} \qquad (4.17)$$

which leads to a signal current

$$I_S = V_S/R$$

R being the thermopile resistance.

$$I_S = \mathscr{S} \psi \cos(2\pi ft - \gamma)/R\{(1/\mathscr{R}^2) + (2\pi f \, kAs)^2\}^{1/2}. \qquad (4.18)$$

The thermal capacity of the junctions, kAs, is kept low in contemporary instruments by employing thin, conducting films, evaporated onto a membrane rather than bulk wires or strip, to form the pile. The value of $\mathscr{S}$ will be some 10^{-4} volt per degree of temperature difference, this being the figure for an antimony–bismuth couple. Some piles use multielement alloys of copper–silver–tellurium and selenium to obtain large $\mathscr{S}$ and these alloys are probably the results of painstaking empirical research. The value of the heat link $\mathscr{R}$ is high with the membrane construction discussed, especially if the couple-bearing membrane is placed inside an evacuated enclosure. However, the unavoidable heat conductance is that down the thermoelectric materials themselves and, although this can be made low by using long, small cross-section thermocouple arms, these very factors will make the electrical resistance large. As the conflicting requisites of low thermal and high electrical conductance cannot then be met by design geometry it is necessary to use materials showing large ratios of σ, the electrical conductivity, to K, the thermal conductivity.

In practice radiation powers of 5×10^{-11} watt may be detected with a good thermopile of evaporated-film construction, and the response time of $\sim 10^{-5}$ s or a value for f of $\sim 10^4$ Hz. The spectral range which can be covered is similar to that for the bolometer, i.e. up to some hundreds of microns wavelength.

4.6. Pyroelectric detector

The newest member of the thermal group that will be discussed at any length, is the pyroelectric detector. Some crystal structures have a symmetry, or, more properly, lack of symmetry, which results in their showing permanent electrical charges on certain crystal faces. These charges appear in equal and opposite quantities on parallel faces and so constitute dipoles. In practice a crystal cut to expose these dipole faces would not, indefinitely, show charge, as small oppositely charged particles would be attracted and eventually neutralize the dipole. However, as the magnitude of the charge is dependent on temperature, then, by varying the temperature regularly, and quickly enough, a charge can be produced on the crystal varying more rapidly than any neutralizing particles can adjust themselves (Fig. 4.4). The variation of charge on a unit area of crystal when the

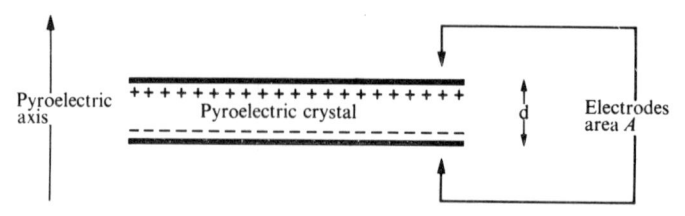

Fig. 4.4. Charges produced by rapid temperature change in pyroelectric detector.

temperature changes by one absolute degree is by definition the pyroelectric coefficient and will be designated Γ. The value of Γ increases as the actual temperature increases until it falls abruptly to zero at a critical temperature—the

Thermal detectors 45

Curie temperature. It is evidently desirable to work near this temperature to enjoy large values of Γ, but not so near that the normal variations in ambient temperature could take the detector past the Curie temperature. A material that has been found eminently satisfactory for its pyroelectric constant, Curie temperature, and thermal parameters is triglycine sulphate (T.G.S.).

We must now consider in detail the electrical effects that temperature changes cause in the circuit associated with a slice of pyroelectric material being used as a detector. The most common form of construction is basically a thin slice of material cut from the crystal so that the large-area faces display the generated charges. These faces are coated with thin electrically conducting films to form the contacts for the detector and the radiation to be observed is incident on either one of them. If the detector area is A then a charge of $\Gamma \Delta T A$ will appear at the opposite faces of the crystal slice when the temperature is changed by ΔT. If the slice thickness is d then the electrical capacity can be obtained as

$$\epsilon A/d \qquad (4.19)$$

treating it, in effect, as a parallel plate capacitor. Here ϵ is the dielectric constant of the pyroelectric material. From these charge and capacity expressions a signal voltage can be obtained

$$V_S = \text{charge/capacity} = \Delta T \Gamma d/\epsilon. \qquad (4.20)$$

The first hint of non-conformity in this detector is seen when the short-circuit signal current is sought, for now the available charge $\Delta T A \Gamma$ must pass through the external circuit in a time equal to the assumed interruption period of the radiation, i.e. $1/f$ seconds, and the mean signal current I_S is therefore

$$I_S = \Delta T A \Gamma / (1/f) = \Delta T A \Gamma f. \qquad (4.21)$$

Substituting the standard expression for ΔT which is also frequency dependent gives

$$I_S = \left[\psi \cos(2\pi ft - \gamma)/\{(1/\Re^2) + (dAs\, 2\pi f)^2\}^{1/2} \right] A \Gamma f. \qquad (4.22)$$

Surprisingly this signal current increases with f until saturation sets in at frequencies substantially greater than $1/\Re\, dAs\, 2\pi$. The variation of V_S, the signal voltage, with f is complementary to that of I_S, having a large steady value at low frequencies falling inversely with the frequency at higher values (Fig. 4.5).

$$V_S = \left[\psi \cos(2\pi ft - \gamma)/\{(1/\Re^2) + (dAs\, 2\pi f)^2\}^{1/2} \right] \frac{\Gamma d}{\epsilon}. \qquad (4.23)$$

The high-frequency amplitudes of I_S and V_S can be approximated:

$$|I_S|_{f \to \infty} = \psi \Gamma / 2\pi d\sigma, \qquad (4.24)$$

$$|V_S|_{f \to \infty} = \psi \Gamma / 2\pi f A s \epsilon. \qquad (4.25)$$

46   Thermal detectors

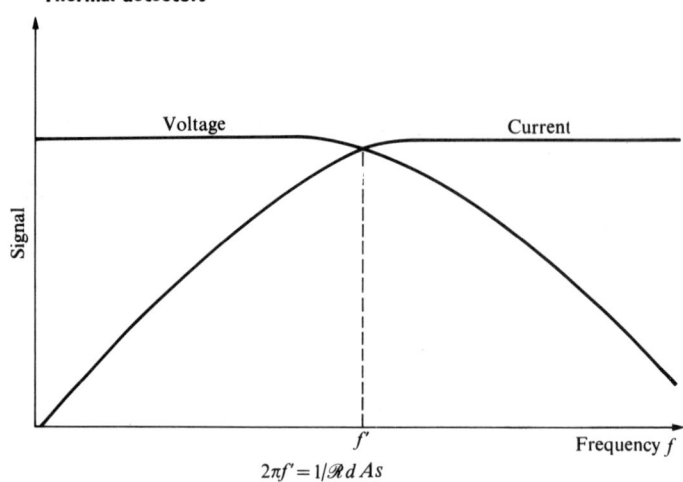

Fig. 4.5. Form of pyroelectric-signal variation with radiation interruption frequency.

The minimum radiated power that can be detected with a good pyroelectric device operating at room temperature will be about 5×10^{-10} watt working with an interruption frequency of 10^3 Hz. The spectral range extends out to thousands of microns wavelength. It is not limited to wavelengths which the pyroelectric material will absorb as the electrode material itself may be used to absorb the radiation and then conduct the thermal energy to the crystal.

The performance parameters for thermal detectors are collected for reference in Table 4.1.

TABLE 4.1.

Device	Current signal	Voltage signal	Response time
Pneumatic (Golay)	$\dfrac{KA\psi \cos(2\pi ft - \gamma)}{2\pi ft \,[\alpha V + A^2 P]}$		Mechanical limitation $\sim 10^{-3}$ s.
Bolometer		$\dfrac{IR\beta\psi \cos(2\pi ft - \gamma)}{[1/\mathcal{R}^2 + (hAs\, 2\pi f)^2]^{1/2}}$	Decreasing temperature amplitude limits to $\sim 10^{-3}$ s.
Thermopile	$\dfrac{\mathcal{S}\psi \cos(2\pi ft - \gamma)}{R\,[1/\mathcal{R}^2 + (WAs\, 2\pi f)^2]^{1/2}}$	$\dfrac{\mathcal{S}\psi \cos(2\pi ft - \gamma)}{[1/\mathcal{R}^2 + (hAs\, 2\pi f)^2]^{1/2}}$	As above to 10^{-5} s.
Pyroelectric	$\dfrac{\psi \cos(2\pi ft - \gamma)}{[1/\mathcal{R}^2 + (dAs\, 2\pi f)^2]^{1/2}} A\Gamma f$	$\dfrac{\psi \cos(2\pi ft - \gamma)}{[1/\mathcal{R}^2 + (dAs\, 2\pi f)^2]^{1/2}} \Gamma\dfrac{d}{\epsilon}$	Complex response with frequency. Normally used $10-10^3$ Hz.

5 Less orthodox detectors

5.1. The unconventional detector and indicator

The devices and systems described in this chapter cannot be fitted into the two major families of thermal and electronic detectors that have been examined so far and so we will content ourselves with qualitative descriptions rather than spend much time on a number of analyses each applicable to only one detector. Three of the detector systems to be considered, The Evaporagraph, The Absorption-Edge Image Converter, and The Liquid Crystal Detector are used only as indicators of the spatial pattern of infrared radiation rather than as instruments for the precise measurement of radiation intensity. The remaining five systems can be made to yield quantitative signals.

5.2. The evaporagraph or eidophor

The evaporagraph, first prepared by Professor Fischer in late 1948 is used to convert a pattern of infrared radiation into a visible picture without the intermediaries of detector, amplifier, and display system. Essentially the instrument consists of a sealed chamber separated into two parts by a plane membrane. The space on one side of the membrane contains oil-vapour while that on the other side is evacuated. The infrared image is focused onto the side of the membrane facing into the evacuated space and causes a temperature pattern corresponding to the infrared image to be formed. Those parts of the membrane at higher temperatures will allow less oil-vapour condensation than other, cooler, regions and so a pattern of oil deposition builds up reproducing the radiation pattern, and, if the oil is illuminated with white light, then interference colours depending on the oil thickness, will make the pattern visible (Fig. 5.1). Modifications of this basic idea yielding faster, more sensitive systems have been prepared and examined practically. Those variations have been mainly concerned with increasing the deformation of the oil film on the membrane and observing this deformation more sensitively. A sophisticated technique reported by Mast and La Roche places an opaque grid in the radiation pattern. Corresponding to the windows in the grid will be a pattern of comparatively isolated regions of higher-temperature oil in which the surface tension is lower than that of the surrounding oil and, because of this, these areas deform into dimple-like shapes. Clearly where no radiation is falling there will be no temperature variation between the oil under the window

48   **Less orthodox detectors**

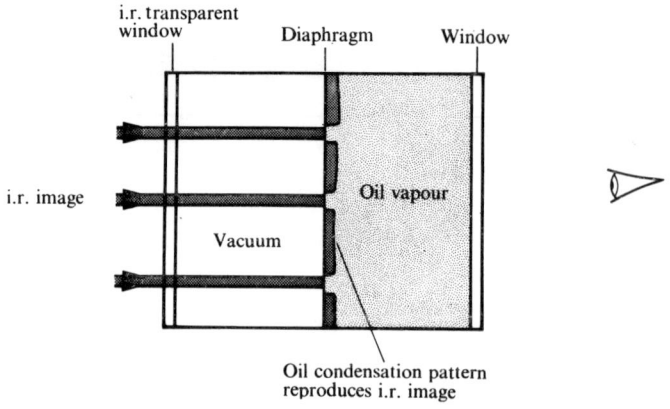

Fig. 5.1. Basic evaporagraph.

and that under the bars of the grid, therefore the surface will remain flat. This surface pattern is observed by a 'Schlieren' system wherein the oil film is illuminated and light reflected from it observed, using an optical system which obscures all light save that coming from regular arrays of dimples, where they exist.

Although the response of this type of instrument is necessarily slow, it produces a good optical image with a mechanically simple and robust apparatus.

5.3. The absorption-edge image converter

An ingenious concept, that for some reason has not enjoyed widespread use, exploits one of the optical properties of semiconductors mentioned in Chapter 2. The transmission of visible radiation in a semiconductor is usually large at long wavelengths and falls abruptly, at a certain critical wavelength λc, to a low value which is maintained for all shorter wavelengths. The value of λc is temperature dependent, usually becoming longer with increasing temperature (Fig. 5.2). Suppose now a plate of such material has light of wavelength just longer than λc falling on it, then most of this light will be transmitted through the plate. If

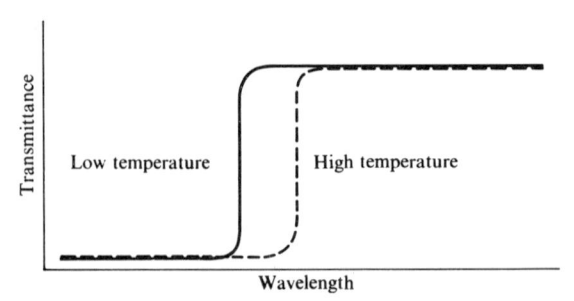

Fig. 5.2. Variation in semiconductor fundamental absorption wavelength with temperature.

Less orthodox detectors

now infrared radiation falls on the element the temperature will rise, λc will increase, and light will no longer be transmitted. In practice this absorption-edge image converter has been made by supporting a thin film of selenium, in an evacuated enclosure to avoid convection currents, and focusing the image of an infrared scene onto it through a suitably transmitting window. The heat pattern, reproduced as an opacity pattern in the selenium, was made visible by shining monochromatic sodium light through the film, so giving a 'negative' image of the infrared pattern, areas of intense radiation appearing dark and vice versa (Fig. 5.3). The device yielded recognisable pictures of a kettle at $60°C$ and a human head, presumably at a lower temperature. The response speed was about 0·5 s.

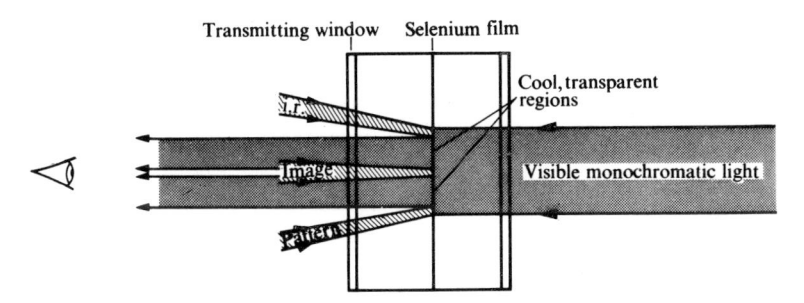

Fig. 5.3. Schematic of absorption-edge image converter.

5.4. Liquid crystals

Materials designated as liquid crystals (L.C.) show an apparent colour which is a sensitive function of temperature, a good L.C. changing its hue by 1000Å for a temperature change of one degree and taking a few tenths of a second to do so. The formation of visible images from thermal patterns is an obvious exploitation of these L.C. properties, but before discussing their use we will examine the mechanism which effects these apparent colour changes.

A group of organic chemical compounds derived from cholesterol, a large carbon + hydrogen molecule, shows an intermediate stage between the low-temperature, completely crystalline, state and the wholly amorphous liquid form which occurs at high temperatures. This intermediate stage is made up from layers of molecules, all those in any one layer pointing in the same direction and so defining a layer direction. This direction changes regularly from layer to layer so that after a certain number of layers it will have rotated through $360°$ and so defined a pitch for the L.C. If white light with random polarization is incident on this structure one wavelength will be very efficiently transmitted giving the crystal a dominant colour. This favoured wavelength depends on the pitch of the L.C. structure, which in its turn depends on the temperature, and so we have the required variation of colour with temperature.

Most materials will change colour if the temperature varies strongly enough

50 Less orthodox detectors

but few will match the 1000Å per degree shown by L.C.s and even fewer will be able to reverse their colour changes as the temperature fluctuates.

The working temperatures for L.C.s is conveniently around room temperature and much of their use has so far been in the direct observation of an L.C. film painted over a machine or component in order to show up the temperature distribution. It is arguable that this form of thermal imaging is hardly part of infrared practice but, although use so far has been by direct thermal contact, there seems no reason why an infrared image should not be made visible by focusing onto a large-area L.C.

5.5. Microwave-biassed photoconductors

It will be remembered, from the discussion of the photoconductive effect, that the difficulty in increasing the signal by increasing the voltage applied to the sample arose from the free-carrier recombination at the contacts. A higher electrical field merely drove holes and electrons more quickly into close proximity at the contacted ends of the device where they rapidly recombined and quenched the signal. The cure for this was to add impurities which formed traps immobilizing either all the free holes or all the free electrons: the penalty associated with this cure was the slow discharge of carriers from the traps giving a current long after the radiation signal had ceased.

The dilemma of choosing speed or sensitivity was resolved by Sommers and Teutsch (two workers at the R.C.A. laboratories) who suggested that, even without the one-carrier traps, excessive recombination at the contacts could be avoided if the electric field being applied were reversed just before the cloud of photo-excited carriers reached a contact. The velocity of the excited carriers and the detector dimensions required the field reversal to take place some 10^{10} times a second which is a frequency in the microwave region. Now microwaves, that is electromagnetic radiation having a wavelength falling between a few millimetres and tens of centimetres, do not require the conventional circuitry of low-frequency or direct-current practice, with wires attached to contact regions on the photo-conductor specimen. It is possible to pass power directly into the specimen merely by holding it in a microwave beam which will produce within it the high-frequency field required. (This cavalier simplification of the coupling of specimen and microwaves is perpetrated only to make explicit the extra practical benefit that the technique produced in removing the necessity for contacts on the specimen.) There is a limitation on the microwave-biassed (M.W.B.) photoconduction system which is imposed by the rapidity with which the free-carrier population in the detector can respond to external field reversals. This response rate is controlled by the dielectric relaxation time of the material, equal to the ratio of the permittivity to the electrical conductivity.

In practice a beam of microwaves is led through a tube or guide into a resonant chamber where the photoconductive specimen is held in a position of maximum electric field. The currents set up in the specimen absorb power from the microwave beam and the remaining power in the beam is measured at the exit of the chamber. If the specimen is now irradiated, then the consequent fall in conductivity will cause it to absorb even more microwave power and the through-put of the resonant chamber will decrease (Fig. 5.4). It is this variation that is used to observe the presence of infrared radiation on the detector.

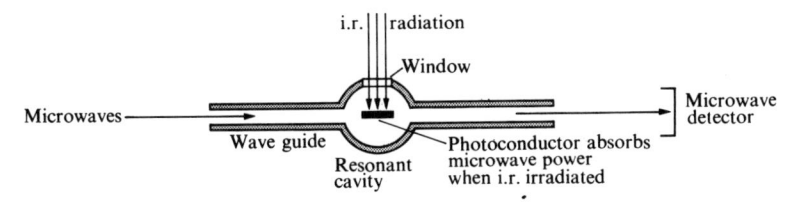

Fig. 5.4. Schematic of microwave-bias photodetection.

This system of microwave guide and resonant chamber also acts in the manner of an electrical transformer increasing the voltage that finally appears in the microwave-detector circuit, over that appearing across the photoconductor itself, by a factor of ten or more in practical systems.

This system has been most intensively developed using room-temperature photoconducting material sensitive to 1·5 μm radiation, but 10 μm-wavelength radiation has been detected by a system operating at very low temperatures.

5.6. The Josephson-junction detector

The Josephson junction consists, in essence, of two superconducting electrodes separated by some 10Å of insulator. Pairs of electrons can tunnel from electrode to electrode through the insulator without doing work. This means, macroscopically, that a finite current can flow across the junction without an applied voltage difference and it behaves as if it has no electrical resistivity. If the current through the junction is increased beyond a certain critical value then the bond between the electron pair fails and the junction behaves as a simple ohmic resistance. If now electromagnetic radiation of frequency f falls on the junction then the current–voltage relationship will show a normal ohmic resistance until a voltage V given by $hf = 2Vq$ is reached when the bond reforms. (Here h is Planck's constant.) At this voltage a sudden increase in current appears giving another region of zero resistivity (Fig. 5.5). It is now clear that by monitoring the appropriate region of the current–voltage characteristic shown by the junction the presence of incident radiation can be detected. Although the Josephson junction was first used for millimetric wavelengths it seems possible that it may be used to detect radiation with a wavelength as short as 200 μm. The device response time is less than 10^{-9} second and will probably be limited always by the time constant of the electrical circuit in which it is used. Radiation with a power of less than 10^{-11} watt has been detected

52   Less orthodox detectors

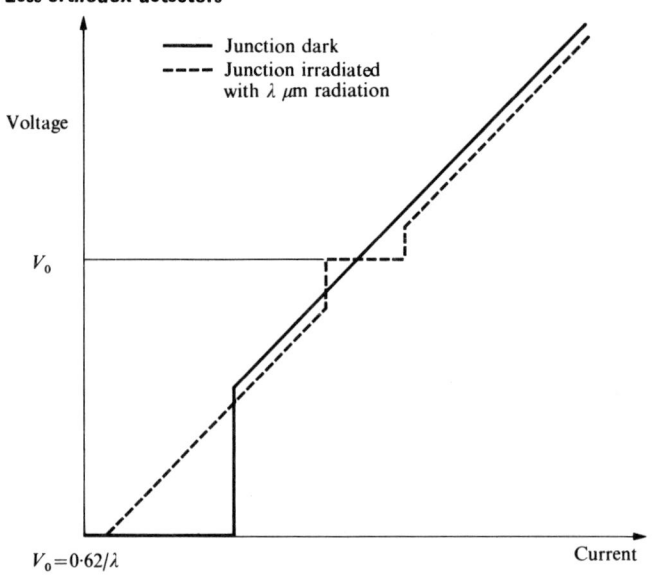

Fig. 5.5. Current–voltage characteristics of Josephson junction.

by a junction using niobium superconducting arms held at a temperature of 1·3K. This operating temperature draws attention to the practical disadvantage in using what is a fast and sensitive infrared detector: the necessary superconducting state is achieved only at very low temperatures.

5.7. The electronic bolometer

The mechanism of photoconductivity that has been both explicitly and implicity assumed so far in this book is one in which the effect stems from a change in number of free charge-carriers, the change being caused by production of mobile holes or electrons or both from radiation energy. Now the conductance of the material depends equally on the mobility of available charge-carriers, i.e. on their drift velocity under the influence of unit electric field. Putley of the Royal Radar Establishment discovered in 1960 that incident radiation of long wavelength increased the mobility of electrons in indium antimonide held at a temperature of about 4K. The mechanism of this mobility increase is best understood by regarding the carriers as an electron gas which is heated by the radiation, the electrons becoming more mobile as they grow hotter. The scale of the effect is such that it can only be seen at very low temperatures since at normal temperatures the heated carriers would very rapidly give up their thermal energy to their crystalline host which, having so great a thermal capacity, would be relatively unaffected. The 'electronic bolometer' which can be made by exploiting this effect is capable

Less orthodox detectors

of detecting radiation with a power as small as 10^{-12} watt and having a wavelength in the range 300 μm to 1 cm, this broad band sensitivity being, of course, typical of a thermal detector.

5.8. Magnetically tuned infrared detector

Another long-wavelength detector using indium antimonide at very low temperatures works on a more conventional basis in which low-energy photons free carriers from impurity levels which are, energetically, close to the conduction band. This sounds so standard a mechanism as to be out of place in a chapter devoted to the less orthodox device but the interest lies in the reconciliation of sensitivity and wavelength response. If only a few impurity centres are added to the compound then the probability of exciting carriers from them with any given radiation flux is low. Now by increasing the number of centres this probability becomes larger and it would be expected that the sensitivity would do so as well but, unfortunately, the energy level of the centres now begins to spread and may finally reach right to the conduction band. The desired situation with a discrete, well-defined impurity level can be restored by the presence of a high magnetic field leading to a sensitive long-wavelength, or small photon-energy, detector. The effect of the magnetic field is not only to compress the band of energies that the impurities show but also to move it slightly with relation to the conduction band (Fig. 5.6). This allows different wavelength radiations to be preferentially

(a) Few impurities, photodetection is inefficient

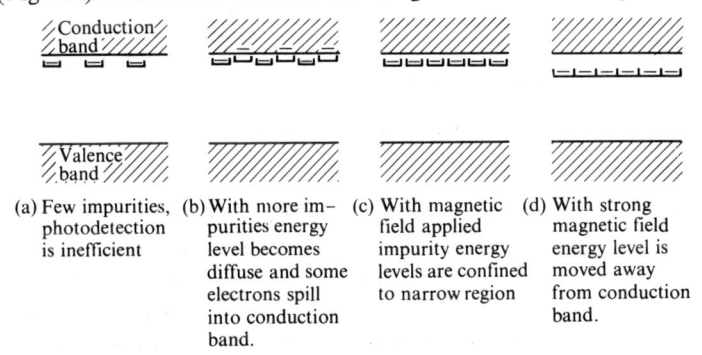

(b) With more impurities energy level becomes diffuse and some electrons spill into conduction band.

(c) With magnetic field applied impurity energy levels are confined to narrow region

(d) With strong magnetic field energy level is moved away from conduction band.

Fig. 5.6. Use of impurity centres in infrared photodetection.

detected as the magnetic field is varied and so the detector is said to be tunable. One example of an indium-antimonide tunable detector showed a peak response at a wavelength varying from 150 μm with a magnetic field of 10 kG to 25 μm when 75 kG was applied. Radiation intensities of about 10^{-11} watt can be detected with a response time of some 10^{-7} second.

5.9. The photon-drag detector

The momentum from a flux of photons can be used to drive free carriers down to one end of a semiconducting bar, with the consequent charge inhomogeneity

54   Less orthodox detectors

appearing as an external voltage-gradient along the semiconductor in the direction of the incident radiation (Fig. 5.7). This effect is not to be confused with simple photoconduction where the energy of the photon frees a charge carrier for subsequent manipulation by an external electric field: here it is the momentum of

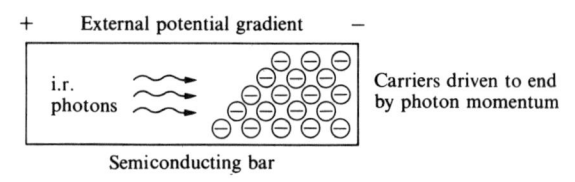

Fig. 5.7. Photon-drag detector.

the photon that pushes along carriers that are already free. Now the momentum p of a photon having an energy E is given by the relationship

$$p = E/c \tag{5.1}$$

c being the velocity of light. If P such photons per second bombard N free electrons then the force acting on the electrons will be PE/c by Newton's second law. In equilibrium this must be opposed by the electric field $\mathcal{E}$ set up in the semiconductor and so

$$N\mathcal{E}q = PE/c \tag{5.2}$$

$$\mathcal{E} = PE/cNq \tag{5.3}$$

The signal voltage appearing along a detector of length L is given by

$$V_S = \mathcal{E}L = PEL/cNq. \tag{5.4}$$

The sensitivity of these 'photon-drag' detectors is low, since the momentum associated with a photon is very small even when its energy is reasonably large. It may appear perverse to use momentum effects when the much larger energetic phenomena are available but the response speed of the photon-drag detector is very great and probably incommensurable since, in any complete detector system, some other component will almost certainly be slower and hence form the limiting factor. This fast, insensitive device is used chiefly for monitoring the output of lasers operated in very brief pulses as, in this situation, much radiation power is available. Commercial devices develop a signal of a few microvolts for an incident radiation power of a watt and respond in less than 10^{-9} second.

5.10. M.O.M. detector

The last device to be considered has been used to detect radiation with wavelengths lying in the 10–20 μm range to which it responds in less than 10^{-13} second. Designated as M.O.M., or metal–oxide–metal detector, the construction is simple, at least to describe, consisting typically of a short, pointed tungsten wire pressed against a nickel or steel block. The length of the tungsten does not seem to be critical, 2–3 mm being satisfactory, but the diameter of the probe must be ~ 10 per cent of the radiation wavelength, say 2 μm. This probe probably resonates with the radiation and the metal–oxide–metal contact at the nickel rectifies the consequent induced currents. In fact the metal probe may be regarded as a very high-frequency dipole receiving aerial. The normal mode of use is to place the probe assembly in a microwave beam so that when the infrared radiation falls on it, the excited condition of the device affects the microwave propagation.

6 Noise, signal processing and coefficients of performance

6.1. General aspects of noise and signal processing

The final result that we seek from any radiation-detection system is a sensible indication of the presence and intensity of the radiation. This can be visual, using the displacement of a needle on a meter scale or the trace on an oscilloscope; it could be aural and employ buzzer or oscillator; indeed the stimulation of any of the senses would serve, although devices using tactile or olfactory effects seem rare.

To be definite, consider a signal which is shown by the position of the pointer on a meter. This will not be absolutely steady but will vary about some mean position. This mean position, estimated over any length of time will, as the observation period is increased, be known with increasing accuracy. If the observation time were increased to infinity then the mean position of the needle, which represents the signal magnitude, would be known exactly, but any finite period of observation will leave a region of uncertainty in the exact needle position. The fluctuations of the needle are the visible consequence of all the 'noise' that the signal has collected in detection and amplification. This noise is a consequence of the random statistical nature of most of the phenomena involved and the several sources will be enumerated and examined in detail later. A significant aspect of noise, manifest here as a needle oscillation, is that it cannot be forecast. Obviously, were the displacements to be in a known pattern then, in principle at any rate, they could be nullified by injecting equal and opposite pulses electrically and they would not then qualify as noise. Another important quality that noise displays is the effective decrease with increasing period of observation which has already been mentioned.

This latter behaviour is significant as it is related to the range of frequencies of varying, or modulated, radiation that can be accepted by the system. This frequency range is called the band-width and if the response time of a detector is τ seconds then it can usefully process frequencies between zero and $1/2\pi\tau$ Hz, i.e. the band-width is $(1/2\pi\tau)-0$ or simply $1/2\pi\tau$ Hz. To exploit this band-width the signal observation period is limited to τ s and, if τ is small, then the signal uncertainty due to noise will be large. Thus a small value of τ gives a large band-width but a poor noise performance, the converse being true for large values of τ.

A convenient unit of magnitude for noise in the system is the signal itself. If the signal is in the form of a voltage then the noise should be expressed as a

Noise, signal processing and coefficients of performance 57

fraction of that signal voltage and analogously for current or power signals. The working maximum for this fraction is normally taken to be unity, that is, a system having noise equal to the signal is regarded as being of as low a sensitivity as is useful. This is, however, only a rule of thumb since any cyclic signal can be retrieved from quite a large noise background using 'box-car' integration or some similarly sophisticated electronics. The basic operation of these signal-retrieval circuits is the repeated sampling of any given part of a cyclic signal until such time as the mean value of the sample fluctuates only within an acceptable amplitude. It will be perceived that this technique is, in fact, exploiting the tenet that says prolonged observation decreases the background noise.

Radiation-detection systems are subject not only to all noise that any sensitive system suffers from but enjoy at least two peculiar to themselves. The first of these is the noise in the signal (N.I.S.) which is basically due to the fact that photons in a radiation stream arrive irregularly and the energy flow only appears to be constant when considered over a prolonged period. The second, background noise (B.N.), is again due to the manner in which a photodetector, or any body, maintains its equilibrium temperature (Fig. 6.1). This is done by the continued loss of photons in radiation to its surroundings and loss of heat by conduction. Each process is, of course, balanced over a long period by photon and heat gain but the photon and heat content will fluctuate above and below the time-mean value.

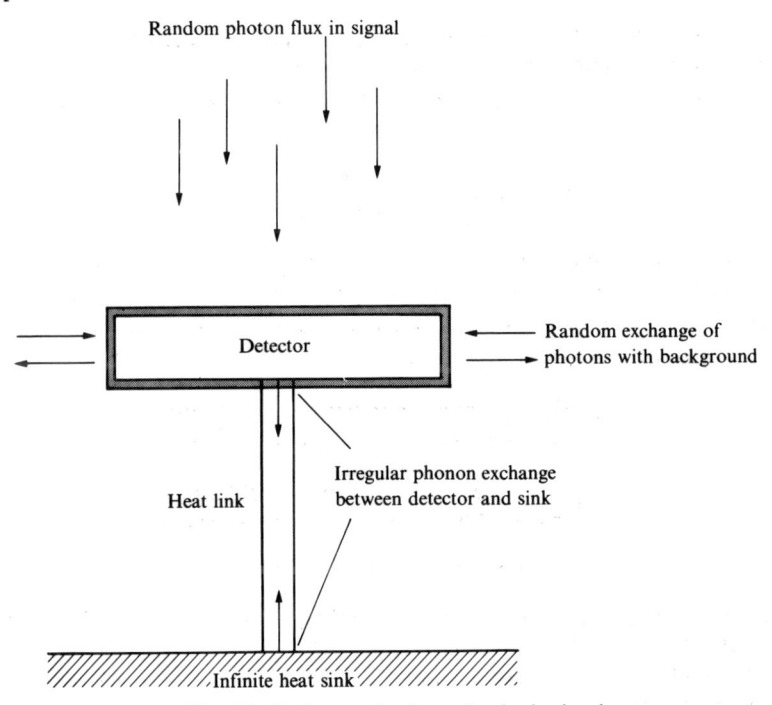

Fig. 6.1. Background noise and noise-in-signal.

Looking now at a real system and its noise sources, we see the signal brings its own noise (N.I.S.) with it. This is joined by (B.N.) at the detector and then the electrical noise in device (E.N.D.).

We now have noise powers, N.I.S., B.N. and E.N.D. which have to be added. (We elect to add the noise powers since this can be done by simple scalar summation for, clearly, total power must be conserved. Adding noise voltages or currents would be more difficult since both of these parameters can vary in a manner depending on the impedance of the component in which they are manifest. This is shown for Johnson noise in section 6.4.) The sum of these noises then goes into the amplifier, is amplified by a factor M, and has yet another noise, the amplifier noise (A.N.), added (Fig. 6.2). What is now reaching the indicating

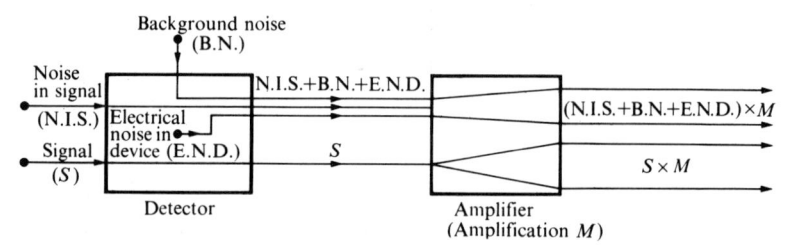

Fig. 6.2. The addition and amplification of noise and signal.

instrument at the end of the system is $(N.I.S. + B.N. + E.N.D.) M + A.N.$ together with a signal S amplified M times. The ratio of the signal to the noise is a guide to the overall behaviour of the system and it is now given by signal to noise ratio,

$$S/N = \frac{S.M.}{(N.I.S. + B.N. + E.N.D.) M + (A.N.)}.$$

It should be noted that although the amplifier has increased the signal available from the radiation by a factor M, this probably being necessary to operate a real display system such as a meter, it has reduced or degraded the signal to noise ratio. This reduction defines the noise figure Q.

$$Q = \frac{\text{signal-to-noise ratio before amplification}}{\text{signal-to-noise ratio after amplification}}$$

$$= \frac{[S/(N.I.S. + B.N. + E.N.D.)]}{[S \times M / \{(N.I.S. + B.N. + E.N.D.) M + A.N.\}]}$$

$$= 1 + A.N./(N.I.S. + B.N. + E.N.D.)M. \qquad (6.1)$$

This has an optimum value of 1 and the amplifier will only degrade the system seriously if $(A.N.) \gg M \times (N.I.S. + B.N. + E.N.D.)$.

The amplifier has so far been assumed to amplify noise and signal indiscriminately. If the signal is in the form of a single frequency oscillation (modulated

Noise, signal processing and coefficients of performance   59

in amplitude to carry information if necessary) the amplifier can be made to operate only for this frequency and so amplify only that part of the entire noise energy which has a similar frequency. The advantages in this are so great that almost all systems use mechanical shutters, colloquially known as choppers, to impress a basic frequency on the radiation, and then an amplifier tuned to this frequency only is employed. In practice the chopper is usually a slotted, or sector, wheel rotated by a synchronous motor with radiation passing through the slots. The more elegant arrangement using slits on an electrically maintained tuning fork is easily envisaged.

It is now evident why the detector anlysis has been developed in terms of a sinusoidally varying signal rather than for a steady flux. Whatever the actual shape of the signal that the interrupter gives, it can be simulated by the sum of a sufficient number of sine or cosine contributions by using Fourier analysis.

The tuned amplifier described above derives its benefit from the rejection of all parts of the total noise offered except for those having the same frequency as the signal. These clearly must not be rejected since they contain the signal. However if, now, all noise is rejected except that having the same phase as well as the same frequency as the signal a further improvement obtains.

To examine this a little more closely, imagine the amplifier to be turned on only when the mechanical shutter is opened to allow the radiation through. When the shutter closes the amplifier is turned off and there is no amplified noise, of any frequency, during that part of the cycle when the radiation is cut off or diminished.

In practice this amplifier switching is done electronically and the system is designated a phase-sensitive-amplifier. Consider the square-wave signal of frequency f shown in Fig. 6.3(a) and let it be subject to the electronically controlled switch, or gate, shown on the same time scale in (b). The signal passes

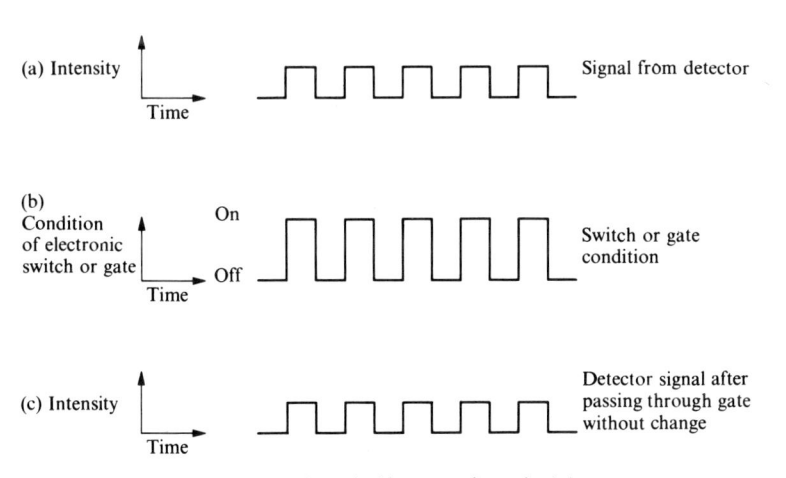

Fig. 6.3. P.S.D. switching or gating principle.

unimpeded through the gate (c) and its mean value can be detected by a D.C. sensitive amplifier. Now a sinusoidal noise of frequency f_n having a phase β with respect to the signal will pass through the gate to produce a noise in the amplifier given by:

noise passing gate = noise strength at time t × degree of switch opening at t.
(6.2)

The variation in the degree of switch opening depicted in (b) is of a square-wave form which can be represented by Fourier analysis as:

$$\text{degree of switch opening} = \frac{1}{2} + \frac{2}{\pi}\left\{\sin 2\pi ft + \frac{1}{3}\sin 6\pi ft + \ldots\right\}$$

Use of the first two terms in this series will be adequate to demonstrate the principle involved and so eqn (6.2) will be written:

$$\text{noise passing gate} = \sin(2\pi f_n t + \beta) \times \left(\frac{1}{2} + \frac{2}{\pi}\sin 2\pi ft\right)$$

$$= \frac{1}{2}\sin(2\pi f_n t + \beta) + \frac{1}{\pi}\cos\{2\pi(f_n + f)t + \beta\}$$

$$- \frac{1}{\pi}\cos[2\pi(f_n - f)t + \beta]. \quad (6.3)$$

This expression has no D.C. component unless $f_n = f$, that is except for noise of the same frequency as the signal. Even this contribution, of magnitude $|\cos\beta|/\pi$, goes to zero as β tends to $\pi/2$, this representing noise which is out of phase with the signal.

Real D.C. amplifiers have of course some response to A.C. signals and hence the noise-produced 'sidebands' at frequencies $(f_n + f)$ and $(f_n - f)$ will be amplified for limited values of f. The larger the response time of the amplifier (or the nearer it is to a true D.C. instrument) the smaller will the accepted sidebands be, for the condition that noise will be amplified is

$$(f_n - f) < 1/2\pi\tau$$

where τ is the amplifier response time. This can be written

$$f_n < f + (1/2\pi\tau). \quad (6.4)$$

Clearly τ should be so large that $1/2\pi\tau$ is negligible compared with f.

A schematic of P.S.D. operation is shown in Fig. 6.4.

The tuned amplifier or the phase-sensitive amplifier must be used if the signal is changing with time but where a repetitive pattern is available then the method of signal recovery known as box-car detection is available. This involves the repeated sampling of the signal at any given part of its cycle and the continued averaging of these samples until all the random noise superimposed on the true

Noise, signal processing and coefficients of performance

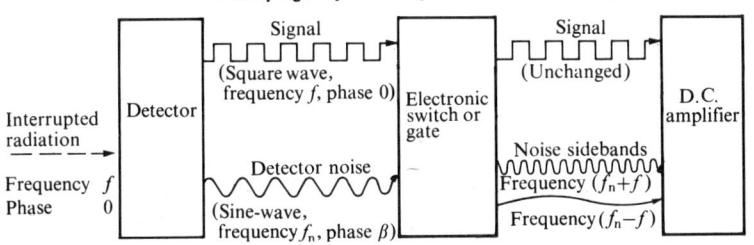

Fig. 6.4. Schematic of phase-sensitive detector.

signal has been averaged to an acceptable minimum. This is a very powerful technique and by its means a signal accompanied by many times its own magnitude of noise can be retrieved if sufficient cycles are sampled and averaged. The name 'box-car' detector comes from the American term for a railway goods wagon and is believed to have been suggested by the resemblance between the repeated silhouettes of a train of these vehicles and the repeated signal shapes offered to the electronic system.

Although the noise-in-signal has been mentioned for the sake of completeness it is not normally a problem since, being proportional to the square root of the signal intensity, it is small except when very strong radiation is being detected.

6.2. Background noise in thermal detectors

The background noise B.N. must be examined more closely as it is a factor in determining the maximum sensitivity that a particular detector could show when working with ideal efficiency. This noise has differing values depending on whether the detector is thermal, photoconductive, or photovoltaic. As we have mentioned, noise stems from statistical fluctuations in 'signal-like' quantities: in this case photons which are being exchanged by the detector with its surroundings to maintain some given temperature. To look at this quantitatively we shall need to use a very general theorem in classical statistics that relates the time-average values of any random number N to the square of its uncertainty $\overline{\Delta N^2}$ by

$$\overline{\Delta N^2} = N. \tag{6.5}$$

It is clear from eqn. (6.5) and Fig. 6.5 that the greater time-average value a number has, the smaller the relative variation it will show. We are concerned with populations of photons, phonons, or free charge-carriers rather than classical particles but assume for the moment that they will obey a similar law. The phonon just mentioned may be considered simply as a packet of heat energy, a concept which allows this population theorem to be used on the heat content of a body.

The procedure that will be followed is to determine the phonon variation in a thermal detector, and hence its temperature fluctuation, and similarly obtain a free-carrier fluctuation for an electronic detector.

Consider the thermal circuit set up on p. 38. If each phonon in the detector is

Noise, signal processing and coefficients of performance

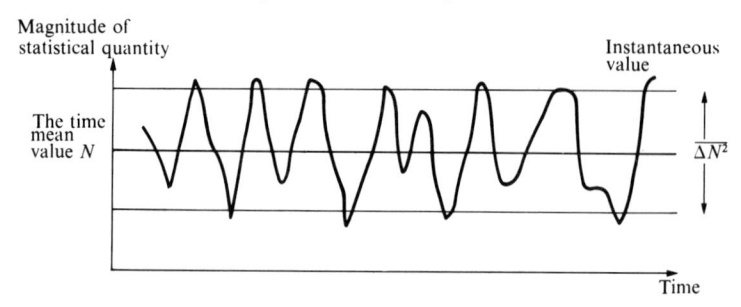

Fig. 6.5. Temporal behaviour of statistical quantity.

assumed to have an energy of kT, then, the number of phonons, N, in the detector must satisfy:

$$\text{heat content} = NkT \qquad (6.6)$$
$$\text{But heat content} = AshT \text{ also,}$$

and so
$$N = \frac{Ash}{k} \qquad (6.7)$$

As $N = \overline{\Delta N^2}$ by the population theorem, the variation in the number of phonons can be written

$$\overline{\Delta N^2} = \frac{Ash}{k}.$$

Since one phonon will give a temperature change of kT/Ash (energy divided by thermal capacity), then the temperature fluctuation $\overline{\Delta T^2}$ will be given by

$$\overline{\Delta T^2} = \overline{\Delta N^2} \times \left(\frac{kT}{Ash}\right)^2 \qquad (6.8)$$

$$= \frac{Ash}{k} \times \left(\frac{kT}{Ash}\right)^2 \qquad (6.9)$$

$$\overline{\Delta T^2} = T^2 \frac{k}{Ash}. \qquad (6.10)$$

So far we have derived a value for the amplitude of these temperature variations but this will be the consequence of random and irregularly changing temperature and can be looked on as composed of an infinite number of pure sine-wave variations of all frequencies (Fig. 6.6). Now a real detector will respond only to stimulation of a suitable frequency. As we have seen, it is limited to a particular range of frequencies or bandpass, and this must be taken into account. Let the detector be connected to an infinite heat sink by a thermal resistance R. The instantaneous noise energy which maintains the dynamic balance between the detector and its surroundings will be written as ψ_t. ψ_t will be assumed to be the sum of a series of sinusoidally varying quantities:

Noise, signal processing and coefficients of performance

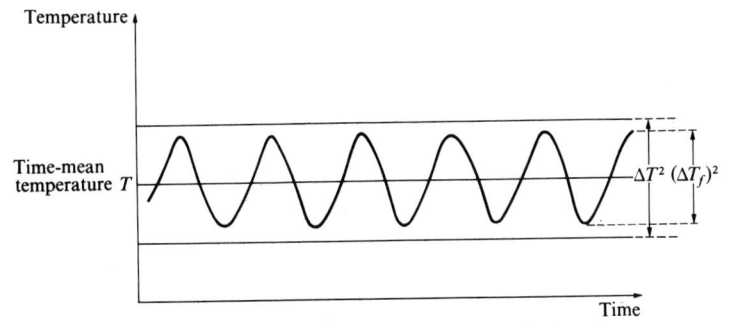

Fig. 6.6. Temporal variation of temperature in detector.

$$\psi_t = |\psi_{f1}| \cos 2\pi f_1 t + |\psi_{f2}| \cos 2\pi f_2 t + \ldots \quad (6.11)$$

The heat balance equation is:

$$Ash \frac{\mathrm{d}}{\mathrm{d}t}(\Delta T) = \psi_t - \frac{\Delta T}{\mathcal{R}}. \quad (6.12)$$

The solution for ΔT may also be written as the sum of the series:

$$\Delta T = |\Delta T_{f1}| \cos 2\pi f_1 t + |\Delta T_{f2}| \cos 2\pi f_2 t + \ldots \quad (6.13)$$

and the amplitude of the general term is:

$$|\Delta T_f| = |\psi_f|/\{(1/\mathcal{R}^2) + (Ash\, 2\pi f)^2\}^{1/2}. \quad (6.14)$$

If the noise is composed equally from oscillations of all frequencies then $|\psi_f|$ is a constant independent of the frequency. Let the constant be $\sqrt{H}$. Then we can write

$$|\Delta T_f|^2 \, \mathrm{d}f = H\mathrm{d}f/\{(1/\mathcal{R}^2) + (Ash\, 2\pi f)^2\}. \quad (6.15)$$

Summing $|\Delta T_f|^2$ for all values of f we obtain $\overline{\Delta T^2}$:

$$\overline{\Delta T^2} = \int_0^\infty H\mathrm{d}f/\{(1/\mathcal{R}^2) + (Ash\, 2\pi f)^2\} \quad (6.16)$$

$$= H\mathcal{R}/4Ash. \quad (6.17)$$

So comparing values of $\overline{\Delta T^2}$ from (6.10) and (6.17) gives a value for H:

$$H = 4kT^2/\mathcal{R}. \quad (6.18)$$

Hence background noise in a band-width $\mathrm{d}f$

$$= H\mathrm{d}f = (4kT^2/\mathcal{R})\,\mathrm{d}f. \quad (6.19)$$

This analysis has been tedious but the result is important as it represents a minimum detectable signal power. To determine the magnitude of this power we must examine the value of the 'heat-link' $\mathcal{R}$ which, by skilful manufacture can be

64 Noise, signal processing and coefficients of performance

made very large, that is, the detector can be almost isolated, thermally, from its surroundings. Even if the device were suspended freely in vacuum however, it would still need to be open to radiation, if it were to function as a detector, and the radiation may be regarded as the last irreducible link. The value of this link is quite easily obtained for, if a detector of area A having a temperature T, is exposed to surroundings with a temperature $T + dT$ (Fig. 6.7) then the net gain

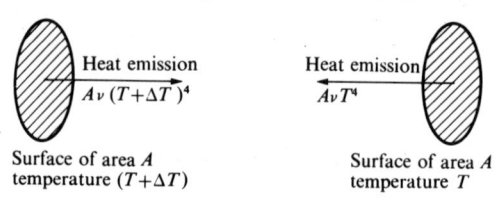

Fig. 6.7. Radiation resistance.

of heat will be $A\nu(T + dT)^4 - A\nu T^4$ if ν is Stephan's constant and the detector surface is of good emissivity. Now the thermal resistance $\mathfrak{R}$ is defined by a relationship

$$\mathfrak{R} = \frac{\text{temperature difference}}{\text{heat flow}} \quad (6.20)$$

$$= \frac{dT}{A\nu(T + dT)^4 - A\nu T^4} \simeq \frac{1}{4\nu A T^3}. \quad (6.21)$$

(There is no point in postulating an emissivity of less than one to maximize this, since lowering emissivity would also decrease the signal absorption.)

We will now put this value of $\mathfrak{R}$ in the expression for the noise power and it becomes

$$16\nu A k T^5 \, df.$$

No improvement can be made on this figure without resorting to cooled shields to intercept the noise radiation from the surroundings in which the detector is used. The uses and limitations of such shields will be discussed generally when coefficients of performance and figures of merit are discussed.

The use of statistical relationships derived for classical particles, such as people, when dealing with photons, phonons, holes, and electrons may be questioned. In fact, the difference between the classical and Bose–Einstein or Fermi statistics will be trivial save in the artificially extreme case of very long-wavelength detectors working at very high temperatures when the full expression,

$$\overline{\Delta N^2} = N / \left\{ 1 - \exp\left(-\frac{1 \cdot 5 \times 10^4}{T \lambda}\right) \right\} \text{ must be used.}$$

In the next section it will be implicitly assumed that an equilibrium population of charge-carriers is maintained by interaction with an environment of photons, despite disparity between the statistical laws governing each type of particle.

Menat has discussed the mechanism of this reconciliation in the work cited in the list of further reading for this chapter.

6.3. Background noise for electronic detectors

Now the analysis of B.N. for electronic detectors will follow a path similar to that leading to the thermal-detector solution. First the flux of photons having energy E to $E + dE$ falling on the detector from its surroundings must be set down. If the detector has a response time τ then the number of photons collected in time τ, and hence the fluctuation in this number, can be calculated. This variation in the collected number is easily translated to a variation in rate of collection. The emission of photons from the detectors will be similarly treated and the consequent noise from these mechanisms obtained.

The Planck radiation law cited on p. 5 in wavelength form can be expressed in terms of photon energy. The flux of photons having an energy between E and $E + dE$ in an environment at a temperature T_E is approximately equal to

$$\delta E \exp(-E/kT_E) \, dE \text{ photons m}^{-2}\text{s}^{-1},$$

where δ is a constant of proportionality. A detector of area A will, in its response time τ, collect N photons where

$$N = \delta A \exp(-E/kT_E) \, dE \times \tau. \tag{6.22}$$

The square of the fluctuation in rate of arrival $\overline{\Delta r^2}$ can now be obtained

$$\overline{\Delta r^2} = \overline{\Delta N^2}/\tau^2 = \delta A \exp(-E/kT_E) \frac{dE}{\tau}. \tag{6.23}$$

The variation in the rate of loss of photons from the detector, $\overline{\Delta l^2}$, is by similar reasoning

$$\overline{\Delta l^2} = \delta A E \exp(-E/kT_D) \frac{dE}{\tau}, \tag{6.24}$$

where the environment temperature T_E has been replaced by the detector temperature T_D.

The effect of the incident photons is to generate free carriers and clearly this has the random but signal-like quality that denotes noise. The loss of photons is accompanied by the loss of charge-carrier pairs since, to emit a photon, a hole and electron must recombine. This is not so evidently a signal-like disturbance, but it is an effective noise in the case of a photoconductive device for, in this, the ancillary circuit will respond without discrimination to both increases and decreases in conductivity. Now a detector can be active, that is be itself a source of power as in the photovoltaic or photo-electro-magnetic detectors; or it can be passive and merely modulate a circuit by resistance changes as in a photoconductive cell. The active detectors are clearly not affected by the photon-loss noise since their ancillary circuits observe voltage signals only, or should do so.

66 Noise, signal processing and coefficients of performance

To determine the power represented by these noise sources, it is necessary only to multiply the rates of photon incidence, or loss, by the energy E of the photons under consideration.

For voltaic devices then

$$\text{noise power} = E\,(\overline{\Delta r^2})^{1/2} \tag{6.25}$$

$$= (\sqrt{\delta})\left[\frac{AE^3 dE}{\tau}\left\{\exp\left(-\frac{E}{kT_E}\right)\right\}\right]^{1/2}. \tag{6.26}$$

For a photoconductive device

$$\text{noise power} = (\sqrt{\delta})\left[\frac{AE^3 dE}{\tau}\left\{\exp\left(-\frac{E}{kT_E}\right) + \exp\left(-\frac{E}{kT_D}\right)\right\}\right]^{1/2}. \tag{6.27}$$

If $T_E \sim T_D$ then this latter expression is $\sqrt{2}$ times as great as that for voltaic devices.

6.4. Johnson or Nyquist noise

Turning to electrical noise, there is one inescapable source in any real device and that is associated with its ohmic resistance. It has been shown by many analysts that any resistance at a temperature T will behave as a noise generator with power $4kT\,\Delta f$ where alternating currents over a frequency range Δf only are considered. If the resistance has a value R then the noise generator must be represented by a voltage source V_n satisfying:

$$\text{power} = \frac{V_n^2}{R} = 4kT\Delta f \tag{6.28}$$

$$\text{or } V_n = 2\sqrt{(RkT\Delta f)}. \tag{6.29}$$

Alternatively it may be regarded as a current generator giving a noise current I_n where

$$\text{power} = I_n^2 R = 4kT\Delta f \tag{6.30}$$

$$\text{or } I_n = 2\sqrt{(kT\Delta f/R)}. \tag{6.31}$$

This is known as Johnson or Nyquist noise in honour of two eminent workers in this field. With most resistances made from metallic or semiconducting materials the value of R is not dependent on the frequency of the applied alternating current, except perhaps at very high values. However, one of our thermal devices, the pyroelectric detector, has an effective resistance which decreases with frequency (Fig. 6.8) and so analyses of noise in that system require the stipulation of a working frequency. The reason for the differing behaviour of the pyroelectric resistance lies in the fact that no continuous current flows through such an insulating material but fixed charges within it are displaced slightly from their

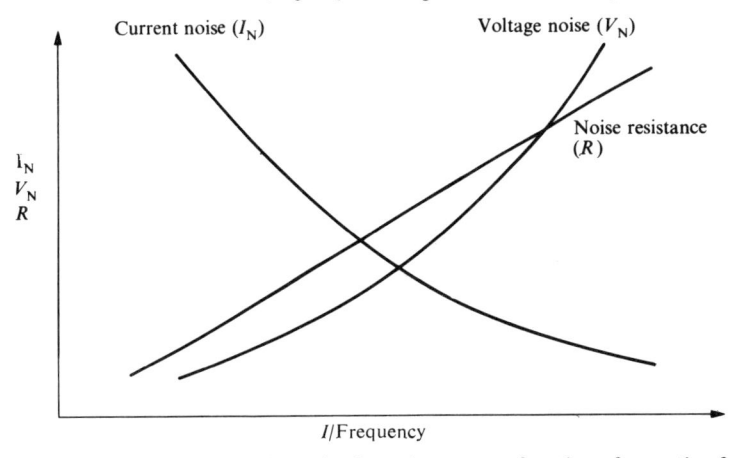

Fig. 6.8. Pyroelectric detector noise and noise resistance as a function of operating frequency.

equilibrium position. This strains the dielectric and such straining absorbs power. This power absorption, which can be equated with an effective resistance, is dependent on the magnitude and frequency of the dielectric strains.

6.5. Current noise

Current noise, sometimes referred to as flicker or $1/f$ noise, varies as the reciprocal of the frequency and the square of the current flowing. If our detector can respond to frequencies lying between f and $f + df$ then the noise current in a circuit with a current I flowing is proportional to $[I^2 \, df/f]^{\frac{1}{2}}$. The mechanism producing this noise is not generally agreed upon but may have its origin in the interfaces between detector and electrical contact material. In concept, at any rate, it can be reduced below any pre-assigned value by decreasing I and therefor, has not the fundamental significance of Johnson noise. Before leaving this point, it should be remarked that the apparently infinite noise current resulting from a zero value of f, can only be had with a zero value of df. This avoids the impossibility by making the noise current indeterminate and accessible only after an infinite period of observation.

Having examined both the signal and the noise separately for various detectors in some detail we can now look briefly at the S/N ratio which appears so frequently in infrared literature.

6.6. The signal-to-noise ratio

With the electronic detectors discussed in Chapter 3, it is usual to consider a perfect but not ideal detector. That is to say, noise which can conceivably be avoided is ignored, but ohmic resistance, the one inevitable source of noise is acknowledged. If the signal is being exploited as a voltage source then the noise must be the voltage associated with the detector resistance R_D and we have

$$(S/N)_{\text{voltage}} = V_S/2\sqrt{(R_D kT\Delta f)}. \tag{6.32}$$

Analogously for current signals

$$(S/N)_{\text{current}} = I_S/2\sqrt{(kT\Delta f/R_D)}. \tag{6.33}$$

The thermal detectors of Chapter 4 follow a similar pattern but the significance of the detector resistance R_D must be made explicit. In the case of bolometers, thermocouples, and thermactors it is simply the electrical resistance of the active element but the Golay cell does not have a meaningful resistance. However, its diaphragm movement is invariably rendered into a sensible signal by some electrical circuit and it is the effective resistance of this circuit which must be used for R_D. It has already been hinted that the pyroelectric detector displays an unusual S/N behaviour which varies with the frequency of modulation, or interruption, in the incident radiation. This is not to be confused with the loss of signal output that any detector exhibits when the frequency in use approaches the response speed of the device, but stems from two particular properties of a pyroelectric detector.

Firstly, the current signal increases its magnitude with frequency. This is because the rate of change of temperature, and therefore the rate of generation of charge, is greatest at the beginning of a cycle of radiation and hence by interrupting the signal frequently this initial, large value is consistently produced. Secondly, the effective resistance of the pyroelectric material decreases with the working frequency as has been already mentioned: in fact it can be written as

$$R_D = K/f$$

where K is some constant of the pyroelectric material and the device geometry. The noise current, as we have seen is

$$2\sqrt{(kT\Delta f/R_D)}$$

and so is proportional to $f^{1/2}$. As both signal and noise currents increase with increasing frequency the S/N ratio enjoys a range of f in which it is practically frequency independent (Fig. 6.9.). This region is, strictly speaking, a very flat, broad maximum in the S/N versus f curve rather than a total independence. An analogous investigation of the signal and noise voltages for a pyroelectric detector shows that the S/N ratio for the voltage mode of use behaves in a similar manner with frequency. The range over which this semi-independence obtains is roughly 10 Hz to 10^3 Hz in a tri-glycine sulphate detector of millimetre dimensions.

6.7. Heterodyne detection

The technique of heterodyning enables the very maximum of response to be extracted from a power-sensitive device. Considerable advantages accrue with this

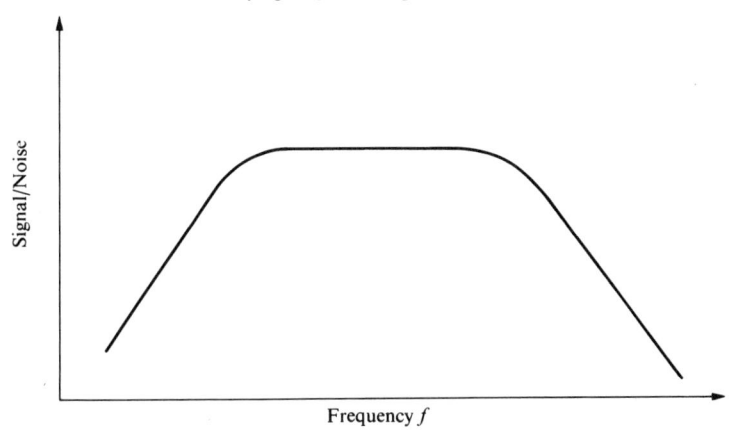

Fig. 6.9. Pyroelectric detector signal/noise variation with operating frequency.

method but the signal to be detected must be in the form of a regular coherent wave train, a condition that would be satisfied by, say, a system using lasers as radiation sources. In principle, the method is exactly the same as that used to process the signal in a superhet radio receiver. Incoming radiation is mixed with locally generated radiation having a very slightly different wavelength and the resultant beat, with the large difference wavelength, is then processed. In the case of the radio this beat wave is at a frequency sufficiently high to be inaudible, hence 'supersonic-heterodyne' contracted to superhet. It is not evident yet that this has produced any advantage but our new signal, although having only the same amplitude variation that the original signal had, now shows a mean amplitude approximately that of the locally generated radiation itself. To make this clearer, let the signal amplitude be A_S and the local radiation amplitude A_R. After heterodyning the signal amplitude varies from $(\frac{1}{2}A_R + \frac{1}{2}A_S)$ to $(\frac{1}{2}A_R - \frac{1}{2}A_S)$ with the beat wavelength. As the power in electromagnetic radiation is proportional to the amplitude of that radiation, then

$$\text{original signal power} \propto (A_S)^2 - (0)^2$$

and

$$\text{after heterodyning} \propto (\tfrac{1}{2}A_R + \tfrac{1}{2}A_S)^2 - (\tfrac{1}{2}A_R - \tfrac{1}{2}A_S)^2,$$

which when evaluated shows a power multiplication of A_R/A_S (Fig. 6.10). However, the noise-in-radiation due to irregular photon arrival, which we previously dismissed, now becomes significant and indeed is the limiting consideration for a heterodyned detector.

The largest value of signal-to-noise ratio that can be obtained with this system is

$$(S/N)_{max} = \sqrt{\left(\frac{\text{radiation signal power}}{E_\lambda/\tau}\right)} \qquad (6.34)$$

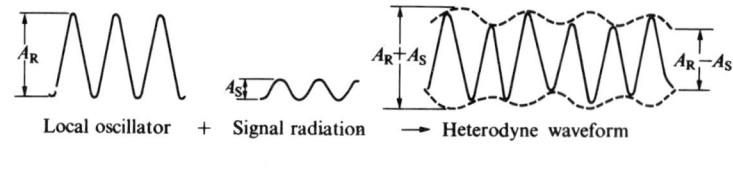

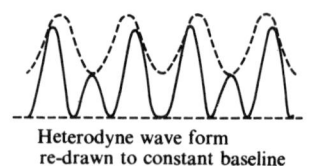

Fig. 6.10. Heterodyning.

where E_λ is the energy of the photons in the local radiation and τ is the response time of the system. It should be noted that this effective noise power of E_λ/τ is not dependent on detector characteristics, other than the response time but is a consequence of the fundamental statistical noise in the postulated local radiation beam.

6.8. Coefficients of performance for detectors

The coefficients of performance that have been proposed for characterizing infrared detectors are many and somewhat subject to fashion. We will look at the most firmly rooted and then see what use they might be to an engineer needing to design an infrared system for actual use.

Responsivity is a very direct measure of the detector's performance being the actual signal voltage or current produced by the device when a certain radiation power is incident on it. The units are volts per watt for voltage responsivity and amperes per watt for current responsivity. This will show the engineer what his amplifier will have to deal with when the magnitude of radiation power or irradiance is known.

Less direct is the Noise Equivalent Power (N.E.P.) and this is the radiation power, normally expressed in watts, which will just give a signal as great as the inherent noise in the detector. Obviously a good device has a small N.E.P. but a problem can arise if the virtue stems from a very small inherent noise rather than a large signal, for then the amplification circuits may be presented with a signal so small that it is swamped by amplifier noise. This possible degradation of available signal-to-noise by the amplifier has already been adumbrated.

A more sophisticated characterization is the detectivity (D^*) which is defined as $\sqrt{(\text{detector area})}/\text{N.E.P.}$ The effect of introducing the detector area into the coefficient is, oddly enough, to make the final value of D^* independent of

Noise, signal processing and coefficients of performance

detector area for many types of device. To demonstrate this, consider a photovoltaic detector used in the current mode.

If P photons s^{-1} m^{-2} fall on a detector of area A generating carriers with an efficiency η then a signal current

$$PA\eta q \text{ amperes}$$

will flow.

The noise current is, in the usual form,

$$2\,(kT\Delta f/R_D)^{1/2} \text{ amperes,}$$

and so a signal-to-noise ratio S/N is obtained where

$$S/N = PA\eta q/(4kT\Delta f/R_D)^{1/2}. \tag{6.35}$$

The photon incidence necessary to make this value of S/N unity is

$$2\,(kT\Delta f/R_D)^{1/2}/\eta q \text{ s}^{-1}.$$

If the photon energy is E_λ then this rate of incidence represents a power

$$E_\lambda\,2\,(kT\Delta f/R_D)^{1/2}/\eta q \text{ W}$$

which by definition is the noise equivalent power N.E.P.

The detectivity D^* can now be written

$$D^* = (\text{detector area})^{1/2}/\text{N.E.P.} \tag{6.36}$$

$$= \eta q\,(AR_D)^{1/2}/E_\lambda\,2\,(kT\Delta f)^{1/2}. \tag{6.37}$$

Now the detector resistance R_D will be proportional to $1/A$, if the elecrodes are plane and parallel with the irradiated detector surface, and so the factor $(AR_D)^{1/2}$ will be constant, dependent only on the detector material. Most photodetectors do show the electrode geometry required and so their detectivities are criteria for the material and processing used, unobscured by differences in size which they may exhibit.

Before the detectivity can be quoted for any device, a number of conditions of operation must be stipulated. First the radiation considered may be either monochromatic, of an appropriate wavelength to excite maximum response, or it can be the total radiation from a black body of specified temperature T. These two conditions are denoted by subscripts λ or T respectively giving D^*_λ, or D_T. Now f, the frequency of interruption of the radiation, and the band pass Δf must be specified. By the time the detectivity symbol reaches the literature it will look like this:

$$D^*_\lambda\,(\lambda, f, \Delta f) \text{ for monochromatic radiation}$$

or

$$D^*_T\,(T, f, \Delta f) \text{ for total black-body radiation.}$$

72   Noise, signal processing and coefficients of performance

The units of detectivity are W and a typical value for a 10 μm detector used at 10^3 Hz with a 1 Hz bandpass would be written

$$D^*_\lambda (10\mu m, 10^3, 1) = 10^6$$

As the detectivity is improved, the danger of the 'lowest detectable' signal power becoming comparable with the background noise arises. When these two are equal then further increases of detectivity can only be exploited in the special conditions mentioned below and the limiting detectivity which allows a signal just equal to background radiation to be detected is referred to as the Background Limit for Infrared Photocells (B.L.I.P.). The B.L.I.P. is arbitrary to the extent that, as we have seen, information can be extracted from a signal plus noise mixture even when S/N is less than unity, and, furthermore, the background radiation can be decreased by putting filters before the detector which will only allow the required wavelengths of radiation to pass. These filters must, of course, be cooled sufficiently to prevent them acting as 'hot' black-body sources of radiation and negating their filter effect.

A measure of the value of these coefficients of performance can be made by attempting to specify a detector for use with a particular signal. The first questions that must be answered are:

(a)   What is the wavelength or wavelength range of the radiation?
(b)   What is the power of the radiation?
(c)   What is the area of the radiation, after any imaging that may take place in the system?
(d)   What is the frequency of variation of the signal, that is the modulation which carries the information?
(e)   Is the signal repetitive or continuously varying?

The answer to question (a) clearly limits the photocells which can be considered and that to question (b) will eliminate devices with N.E.P. values greater than this. Ideally the answer to question (c) should equal the detector area since if that of the imaged radiation is greater then some will be wasted, if less then the unused area of the detector will be a source of noise without signal. The frequency of cell response must be compatible with the modulation rate and so the answer to question (d) will narrow the field even more while the response to (e) may cause a reconsideration of the N.E.P. value that was deemed acceptable at (b) since a repetitive radiation signal can have less power than the N.E.P. of the detector if sophisticated signal retrieval is practised.

These, it was said, are only the first questions that must be examined and the system engineer will need to consider cooling requirements, the effects of possible vibration, mechanical mounting problems, and so on. Only after having produced the right answers to all the questions and then constructed the system accordingly, can the real criterion be used. Does it work?

Further reading

Chapter 1

Cross, A.D. *and* Jones, R.A. (1969). *Introduction to practical infrared spectroscopy.* Butterworth, London.

Dryer, J. R. (1965). *Applications of absorption spectroscopy of organic compounds.* Prentice Hall Inc., Englewood Cliffs.

Marsh, J. C. D. *and* Nicolson, I. K. (1971). A review of infrared astronomy. *Astronomy and Space* **1**, 37.

Moss, T. S. (1959). *Optical Properties of semiconductors.* Semiconductor Monographs, Butterworth, London.

Newman, R. C. (September 1969). Infrared absorption due to localized modes of impurity complexes in ionic semiconducting crystals. *Advances in Physics.*

Sidgwick, J. B. (1956). *William Herschel explorer of the heavens.* Faber, London.

Swift, D. W. *and* Thompson, G. V. (September 1972). Seeing in the dark. *The radio and electronic Engineer.*

Applied Optics (September 1968) examines thermal imaging with good pictorial support.

Chapter 2

Bennett, J. M. *and* Ashley, E. J. (February 1965). Infrared reflectance and emittance of silver and gold evaporated in ultrahigh vacuum. *Applied Optics.*

Betts, D. B. (1965). The spectral response of radiation thermopiles. *J. sci. Inst.* **42**, 243.

Billard, P. (1962). Materials for use in infrared optical devices. *Acta Electronica* **6**, 7.

Conn, G. K. T. *and* Avery, D. G. (1960). *Infrared methods.* Academic Press, London.

Hadni, A. (1967). *Essentials of modern physics applied to the study of the infrared.* Pergamon Press, London.

Hauck, J. E. (October 1964). A guide to optical materials. *Materials in design Engineering.*

Heavens, O. S. (1955). *Optical properties of thin solid films.* Butterworth, London.

Jamieson, J. A., McFee, R. H., Plass, G. N., Grube, R. H. *and* Richards, R. G. (1963). *Infrared physics and engineering.* McGraw Hill.

Kimmitt, M. F. (1970). *Far infrared techniques.* Pion, London.

Martin, D. H. (Editor) (1971). *Spectroscopic techniques.* North Holland Publishing Co., Amsterdam.

McCarthy, D. E. (February 1963, April 1965, October 1968). The reflection and transmission of infrared materials. *Applied Optics.*

McCarthy, D. E. (November 1971). Reflection and transmission measurements in the far infrared. *Applied Optics.*

74   Further reading

Parker, P. (1958). *Electronics.* Arnold, London.
Peterson, E. (January 1963). *Optical materials in the infrared.* Research note no. 195, Thomas J. Watson Research Centre, Yorktown Heights.
Ulrich, R. (October 1968). Interference filters for the far infrared. *Applied Optics.*
Wolfe, W. L. (Editor) (1965). *Handbook of military infrared technology.* Office of Naval Research, Dept. of the Navy, Washington D.C.

Chapter 3

Biard, J. R. *and* Shaunfield, W. N. (1967). A model of the avalanche photodiode. *I.E.E.E. Transactions on electronic Devices* **ED 14**, 233.
Kruse, R. W., McGlauchlin, L. D. *and* McQuiston, R. B. (1963). *Elements of infrared technology.* Wiley, London.
Melchior, H., Fisher, M. B. *and* Arams, F. R. (1970). Photodetectors for optical communication systems. *Proc. I.E.E.E.* **58**, 1466.
Putley, E. H. (1966). Solid-state devices for infrared detection. *J. sci. Inst.* **43**, 857.
Shapiro, P. (January 1969). Infrared detector chart outlines, materials and characteristics. *Electronics.*

Chapter 4

Goldsmid, H. J., Savoides, N. *and* Uher, C. (1972). The Nernst effect in Cd_3As_2 – NiAs. *J. Phys.* D**5**, 1352.
Maserjian, J. (December 1967). NASA Technical Brief 67-10505 (deals with thermactor).
Putley, E. H. (1966). See further reading for Chapter 3.
Smith, R. A., Jones, F. E. *and* Chasmar, R. P. (1968). *The detection and measurement of infrared radiation.* Clarendon Press, Oxford.
Washwell, E. R., Hawkins, S. R. *and* Cuff, K. F. (1970). The Nernst detector. *Appl. Phys. Letters* **17**, 164.
Willardson, R. K. *and* Beer, A. C. (Editors) (1970). Semiconductors and semimetals. Vol. 5, *Infrared detectors.* Academic Press, London.

Chapter 5

Gibson, A. F. *and* Kimmett, M. F. (August 1972). Photon-drag detection. *Laser Focus.*
Kruse, R. W., McGlauchlin, L. D. *and* McQuiston, R. B. (1963). *Elements of infrared technology.* Wiley, London.
Applied Optics, June 1965. (Electronic bolometers.)
Applied Optics, September 1968. (Liquid crystals.)
L'Onde electrique, July-August 1968. (Josephson—junction detectors.)
Proceedings of conference on infrared techniques 1971. *Institution of electronic and radio engineers, conference proceeding no.* 22. (Microwave-bias detection and photon-drag detector.)
Proceedings of the electro-optic 1971 international conference, Chicago. 1971 *Industrial and scientific conference,* Management Inc. (The evaporagraph.)
Proceedings of the symposium on submillimetre waves, New York 1970. Polytechnic Institute of Brooklyn. (The MOM electric tunnelling detector.)

Chapter 6

Abernethy, J. D. W. (December 1970). The boxcar detector. *Wireless World.*
Kimmitt, M. F. (1970). *Far infrared techniques.* Pion, London.

Kruse, R. W., McGlauchlin, L. D. *and* McQuiston, R. B. (1963). *Elements of infrared technology.* Wiley, London.
Menat, M. (1970). D* degredation by sky radiation impinging on an infrared detector. *Optics Communications* **1**, 446.
Menat, M. (1970). The continuity of Bose–Einstein photon noise and Fermi electron statistics. *Optics Communications* **1**, 463.
Smith, R. A., Jones, F. E. *and* Chasmar, R. P. (1968). *The detection and measurement of infrared radiation.* Clarendon Press, Oxford.

Index

Absorption, i.r., 11–14
    by atmosphere, 4, 13, 18
    by lattice, 12–13
    on thermal detectors, 20–21
absorption bands, 13
absorption lines, 2
absorption-edge image converter, 48–49
ambipolar diffusion length, 33
amplifier,
    noise, 56–72
    phase-sensitive, 59
    response time, 60
    tuned, 59
analysis of materials, 2
anti-reflection coatings, 11, 13
astronomy, i.r., 1, 4
atmospheric absorption of i.r., 4, 13, 18
avalanche photodiode, 32

background limit for infrared photocell (B.L.I.P.), 72
background noise, 57, 61–66
backward-wave oscillators, 9
band-gap in semiconductors, 11, 26
band-pass filter; *see* filters
band-width, 56
black surface for detectors, 20
black-body radiation, 5–7, 71
bolometer, 36, 41–43, 44, 68
    electronic, 52
Bose-Einstein statistics, 64
box-car integration, 57, 60
Brewster angle, 20

carcinotrons, 9
carrier
    diffusion coefficient, 32
    lifetime, 30, 32, 34, 35
    saturation velocity, 32
    thermal velocity, 33
    transit time, 30
centre-wavelength of filter, 17
Christiannsen filter, 16
communications, 3–4
crystals, liquid, 49–50

Curie temperature, 45

definition of infrared spectrum, 1
detectivity, 70–72
detectors; *see*
    absorption-edge image converter
    electronic bolometer
    electronic detectors
    evaporograph
    Josephson-junction detector
    liquid crystals
    magnetically tuned infrared detector
    M.O.M. detector
    microwave-biassed photoconductors
    photon-drag detector
    thermal detectors
dielectric relaxation time, 50

eidophor; *see* evaporograph
electronic bolometer, 52
electronic detectors, 22–35
    background noise in, 65–66
    see also
    semiconductor photodetector
    vacuum photocell
emissivity, 3, 7, 64
evaporograph, 47–48

Fabry–Perot filter, 16–17
Fermi statistics, 64
filters, 15–20
flicker noise, 67
Fourier analysis, 19, 59, 60
Fourier-transform spectrometer, 19
free-carrier oscillation, 12
frustrated total-internal-reflection, 15
fundamental absorption, 12–13

geological features, detection of, 3
globar filament, 7
Golay detector, 2, 36, 39–41, 42

half-width (filter), 18
Herschel, 1
heterodyne detection, 68–70

78 Index

high-pass filter; *see* filters
hole, positive, 23

image-converter, 23
   absorption-edge, 48–49
incandescent source, 5
iris, 17–18

Johnson noise, 58, 66, 67
Josephson-junction detector, 51–52

Klystron valve, 9–10

lattice absorption, 12–13
lattice-defect resonance, 2
laser, i.r., 4, 7–8, 36, 39
lenses, i.r., 10–14
liquid crystals, 49–50
low-pass filter; *see* filters

magnetically tuned infrared detector, 53
medical diagnosis, 3
mercury arc, 7
Michelson interferometer (used as spectrometer), 18–19
microwave
   generation, 8
   resonant circuit, 17
microwave-bias techniques, 31
microwave-biassed photoconductors, 50–51
mirrors for i.r., 14–15
modulation of current, 9
modulation rate, 72
M.O.M. detector, 54–55
monochromatic radiation, 71
monochromation, 15–20

Nernst effect detector, 38
Nernst filament, 7
noise, 56–72
   amplifier, 58
   background, 57
   background, for electronic detectors, 65–66
   background, in thermal detectors, 61–65
   current, 67
   electrical, in device, 58
   Johnson, 58, 66, 67
   photon-loss, 65
noise current, 66, 67, 68
noise equivalent power, 70, 72
noise figure, 58
noise in the signal, 57, 61
noise power, 58, 64, 66
noise-in-radiation, 69
Nyquist noise; *see* noise, Johnson

opacity to i.r., 4, 10–14
organic analysis, 2

peak power emission for black body, 6
photoconductive detector; *see* semiconductor detector
photo-electro-magnetic detector; *see* semiconductor detector
photomultiplier, 22, 24, 26
photon-drag detector, 53–54
photon-loss noise, 65
photovoltaic detector; *see* semiconductor detector
Planck's radiation law, 5, 65
plasma, 3, 7
   argon, 9
   mercury arc, 7
plasma frequency, 3
polarization in the i.r., 20
polarization of molecules, 12
population theorem, 62
prism monochromator, 18
pyroelectric
   axis, 37
   coefficient, 44
   detector, 44–46, 68
   materials, 37

quantum efficiency,
   photo-electro-magnetic cell, 33
   vacuum photocell, 25

range-finder, 4
RC limit, 25, 30, 32
reflecting-grating instrument, 18
reflection elements, 14–15
reflective polarizer, 20
reflectivity, 10
relaxation time
   of conducting medium, 30
   of dielectric, 50
response time; *see* speed of response
responsivity, 70

Schlieren system, 48
Seebeck coefficient, 43
semiconductor photodetector, 26–35
   photoconductive, 23, 28–31, 32, 34
   photo-electro-magnetic, 24, 33–35
   photovoltaic, 24, 31–33, 34
semiconductors, 2, 8, 48
   absorption of i.r. in, 10–12
   band-gap in, 11, 26
   impurities in, 27, 30, 50
   positive holes in, 23
   surface of, 35
sidebands (D.C. amplifier), 60
signal retrieval circuits, 57

Index

signal-to-noise ratio, 58, 67, 68, 69, 71, 72
solid diode-laser, 8
sources of i.r., 5–10
speed of response, 24, 25, 32, 34, 36, 37, 44, 49, 51, 54
statistics,
   Bose–Einstein, 64
   Fermi, 64
superconductors in detectors, 41, 51–52

T.G.S., 36, 45, 68
thermactor, 36, 68
thermal detectors, 20–21, 36–46
   background noise in, 61–65
   performance parameters, 46:
   see also
   bolometer
   Golay detector
   pyroelectric detector
   thermopile
thermal equation, 38
thermal imaging, 3

thermal resistance, 62–64
thermal time constant, 36
thermionic valve, 9
thermistor, 41
thermocouple, 36, 43, 68
thermoelectric coefficient, 43
thermojunction, 43–44
thermometer, conventional, 2, 36
thermopile, 43–44
thermovoltaic effect, 43
transmission polarizer, 20
transparency in i.r., 10–14
traps for electrons or holes, 30, 35, 50
tuned amplifier, 59

vacuum photocell, 22, 26
velocity modulation, 9

wavelength sensitivity and response, 26, 36, 41, 43, 46, 53
windows for infrared, 10–14
work function, 24